COAL
DESULFURIZATION

COAL
DESULFURIZATION

ROBERT A. MEYERS
Systems Group of TRW Inc.
Redondo Beach, California

with a Foreword by Edward Teller

MARCEL DEKKER, INC.   New York and Basel

6373-0728

CHEMISTRY

Library of Congress Cataloging in Publication Data

Meyers, Robert A   1936-
    Coal desulfurization.

    Includes indexes.
     1.   Coal--Desulfurization.   I.   Title.
TP325.M53   662'.623   77-9928
ISBN 0-8247-6572-9

MARCEL DEKKER, INC.

270 Madison Avenue, New York, New York 10016

Current printing (last digit):
10 9 8 7 6 5 4 3 2

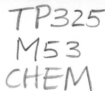

FOREWORD

In the 18th century England ran out of firewood. The British turned to a dirty substitute, coal.

The environmentalists of those days protested but coal was an economic necessity and won out. It probably became the main driving force behind the Industrial Revolution.

Actually, coal has been used in various forms in the production of iron and steel. Indeed, it was needed for this purpose because it gave a hotter flame than wood. The fact that coal was brought into general usage in England more than 200 years ago produced a most important side effect. For the first time in history a truly cheap metal, iron, was available. This is one of the main reasons why the Industrial Revolution could occur.

For a long time coal was king. But in the last few decades it has been more successfully attacked by the environmentalists and dethroned by petroleum. But, in turn, the reign of oil is drawing to a close; one of the main reasons is that the supply appears to be limited. Coal, on the other hand, is plentiful, particularly in the United States. Almost one-third of the world's known coal reserves are located in the United States.

At the same time, the environmental objections are being voiced more strongly than ever and these objections have some justification. In the case of coal, the worst offender is probably sulfur and much of the coal in the United States has a high sulfur content. For this reason, a book on coal desulfurization is of great and obvious importance.

The story is, and should be, long. The foreword, on the other hand, ought to be short. The relevant fact, in a few short words, is that sulfur is present in two different forms; organic sulfur, bound in the organic coal matrix itself, and inorganic, mostly pyritic sulfur dispersed in both the inorganic and organic portion of coal. Part of the pyrite can be separated along with inorganic coal rock and a portion of the organic coal matrix by simple physical means. In some coal deposits, the physically separable sulfur is a big fraction, indeed more than one-half of the sulfur present. In other deposits, the pyrite is not as amenable to physical separation so that a chemical method would be desirable.

It is of the greatest possible importance to understand in a thorough fashion how we can get rid of sulfur in an economic manner. This book addresses this question. A thorough understanding of the problem will make a truly important contribution to the solution of our energy problem.

EDWARD TELLER
Lawrence Livermore Laboratory

PREFACE

Coal is second only to petroleum and natural gas as a source of energy and is slated to surpass the other fossil fuels in production and usage within the near future. Sulfur, contained as inorganic and organic compounds in the coal matrix, provides a serious drawback to the increased utilization of coal. It is my belief that it is technically feasible to remove sulfur from coal by chemical reaction.

This book provides a thorough review of methods for the removal of the sulfur content of coal by chemical means. The review is not simply a summation and correlation of the literature, but rather is highly critical in its treatment. Constant attention is given to process engineering and economic viability. In many cases, published data is recalculated and reinterpreted, giving rise to conclusions differing significantly from those cited in the source literature. In addition, hitherto unpublished work, from our laboratories, is presented for appropriate elucidation of various points.

An attempt is made to systematically categorize methods which can theoretically be used to desulfurize coal. The categorization is based on mechanistic considerations and physical-chemical correlations such as oxidation-reduction potentials, reaction rate constants, solubilities, etc. This has resulted in the definition of a number of potentially rewarding research areas which could lead to new methods for the desulfurization of coal.

To date, one method for chemical desulfurization of coal, ferric sulfate leaching, has been extensively tested. This method is described in detail, starting with the basic chemistry and proceeding through process development and engineering design and cost estimation results. Where applicable, other desulfurization methods reported in the literature are similarly treated. Thus, it is an objective of this book to present the entire spectrum of chemical desulfurization technology — from basic research through engineering for practical application.

A description of the sulfur content distribution of world and U.S. coal and the need for the removal of sulfur from coal are discussed in Chapter 1. The physical and chemical structure of coal as related to coal desulfurization is presented in Chapter 2. The basic chemistry associated with the various possible methods for desulfurization of coal is presented systematically in Chapter 3, while economic considerations are introduced in Chapter 4. Chapter 5 presents a detailed description of ferric sulfate leaching for desulfurization of coal. A number of additional methods are presented in Chapters 6 through 9 for removal of pyritic sulfur, while processes for the removal of organic sulfur from coal are discussed in Chapter 10.

R.A.M.

ACKNOWLEDGMENTS

It is a pleasure to thank the large number of colleagues who contributed directly and indirectly to the preparation of this book.

I wish to thank Mr. B. Dubrow and Dr. J.S. Foster, Jr. of TRW Inc. who encouraged the preparation of this book and provided the use of TRW facilities. I want to express my appreciation for the support of the Environmental Protection Agency in my studies of coal desulfurization, and particularly to Mr. T.K. Janes of that agency for initially proposing the problem to me of the need for coal desulfurization and providing guidance and support during the past five years.

I am very much indebted to Mr. L.J. Van Nice, the brightest scientist with whom it has ever been my pleasure to work, for providing a stimulating technical atmosphere for the preparation of this book, and to his colleagues E.P. Koutsoukos, M.J. Santy, J.W. Hamersma, C.F. Murray, and R.A. Orsini, for providing much of Chapter 5.

My thanks go also to M.V. Melough, S.C. Quinlivan and C.B. de la Fuente who edited and prepared this manuscript for publication.

R.A.M.

CONTENTS

FOREWORD *by Edward Teller* . . . . **iii**

PREFACE . . . . **v**

ACKNOWLEDGMENTS . . . . **vii**

CHAPTER 1. INTRODUCTION 1

  I.   WORLD COAL RESERVES AND SULFUR CONTENT 1
      A. Coal Reserves and Type 1
      B. Sulfur Distribution 3
  II. WHY REMOVE SULFUR? 5
      A. Air Pollution Control 7
      B. Coking Coal 9
  III. WHY CHEMICAL DESULFURIZATION? 11

CHAPTER 2. COAL STRUCTURE AND CHEMICAL
DESULFURIZATION 17

  I.   CHEMICAL STRUCTURE 17
      A. The Organic Coal Matrix 17
      B. Inorganic Coal Structure 19
  II. PHYSICAL STRUCTURE OF COAL 21

CHAPTER 3. CHEMISTRY OF DESULFURIZATION
REACTIONS 25

  I.   CHEMICAL REACTIONS OF PYRITE 26
      A. Classification of Pyrite Reactions 26
      B. Oxidation of Iron Pyrite 27

      1. ELECTROCHEMICAL POTENTIAL 27

      2. CHEMICAL REACTIONS. 28

   C. Reduction of Pyrite 40

II. **DESULFURIZATION OF ORGANIC SULFUR COMPOUNDS** 42

   A. Classification of Desulfurization Methods. 42

      1. SOLVENT PARTITION. 42

      2. THERMAL DECOMPOSITION 43

      3. ACID-BASE NEUTRALIZATION 43

      4. REDUCTION. 43

      5. OXIDATION 44

      6. NUCLEOPHILIC DISPLACEMENT 44

   B. Desulfurization Reactions. 45

      1. SOLVENT PARTITION. 45

      2. THERMAL DECOMPOSITION 47

      3. ACID-BASE NEUTRALIZATION 48

      4. REDUCTION. 48

      5. OXIDATION 49

      6. NUCLEOPHILIC DISPLACEMENT 51

CHAPTER 4. CRITERIA FOR SUCCESSFUL CHEMICAL DESULFURIZATION PROCESSES 55

CHAPTER 5. PYRITIC SULFUR REMOVAL PROCESSES — METAL ION OXIDANTS 59

I. **INTRODUCTION** 59

II. **CHEMISTRY AND PROCESS DATA** 61

   A. Selection of Specific Reagents and Conditions 61

      1. PROCESS DESCRIPTION 61

      2. SELECTION OF SPECIFIC REAGENTS AND CONDITIONS. 62

   B. Rate of Pyritic Sulfur Removal 68

   C. Regeneration of Ferric Sulfate Leach Solution 76

   D. Simultaneous Pyrite Leaching and Ferric Sulfate Regeneration 77

1. SELECTION OF CONDITIONS 77
2. INITIAL ENGINEERING DATA 78
3. ADVANCED ENGINEERING DATA 83

III. RECOVERY OF DESULFURIZATION PRODUCTS 94

A. Removal of Elemental Sulfur from Coal 94

1. VAPORIZATION 94
2. SOLVENT EXTRACTION 97
3. CHEMICAL REACTION 99

B. Rejection of Product Iron and Sulfate 100

C. Fate of Minor and Trace Elements 102

IV. APPLICABILITY TO U.S. COALS 107

A. Sulfur Removal 108

B. Selectivity and Heat Content Changes 109

C. Combination with Coal Cleaning 112

V. ENGINEERING DESIGN AND COST ESTIMATIONS 115

A. TRW Engineering Studies 116

1. SUSPENDABLE COAL PROCESSING
DESIGN AND COST STUDIES 116
2. COARSE COAL PROCESSING DESIGN
AND COST STUDIES 139
3. PROJECTION OF PROCESS ECONOMICS 145

B. Dow Chemical – U.S.A. Design and Cost
Estimation Studies 155

1. PROCESS DESIGN 156
2. PROCESS ECONOMICS 158
3. PROCESS BY-PRODUCTS 161

C. Exxon Research and Engineering Co. Design and
Pollution Control Studies 164

CHAPTER 6. PYRITIC SULFUR REMOVAL PROCESSES –
OXYGEN IN AQUEOUS SOLUTION 175

I. ATMOSPHERIC PRESSURE METHODS 175

II. ELEVATED PRESSURE METHODS 181

A. Process Chemistry and Data 181

B. Process Design 185

III. BACTERIA CATALYZED METHODS 187

CHAPTER 7. PYRITIC SULFUR REMOVAL PROCESSES —
GAS-SOLID METHODS 191

   I.   EXPERIMENTAL DATA 191
  II. ENGINEERING DESIGN AND COST ESTIMATION 197

CHAPTER 8. PYRITIC SULFUR REMOVAL PROCESSES —
CAUSTIC LEACHING 201

   I.   MOLTEN CAUSTIC 201
  II. AQUEOUS CAUSTIC 204

CHAPTER 9. PYRITIC SULFUR REMOVAL PROCESSES —
MISCELLANEOUS OXIDANTS AND REDUCTANTS 211

   I.   SULFUR DIOXIDE 211
  II. NITRIC ACID 215
 III. HYDROGEN PEROXIDE 216
 IV. CHLORINE 220
  V. POTASSIUM NITRATE 220
 VI. REDUCTION 221

CHAPTER 10. ORGANIC SULFUR REMOVAL PROCESSES 223

   I.   SOLVENT EXTRACTION 223
  II. REDUCTION 227
 III. OXIDATION 227
 IV. DISPLACEMENT METHODS 230
      A. Alkali 230
      B. Acid 238
  V. GAS-SOLID PROCESSES 238
   AUTHOR INDEX 241
   SUBJECT INDEX 247

COAL
DESULFURIZATION

CHAPTER 1

INTRODUCTION

   I. World Coal Reserves and Sulfur Content
      A. Coal Reserves and Type
      B. Sulfur Distribution

  II. Why Remove Sulfur?
      A. Air Pollution Control
      B. Coking Coal

 III. Why Chemical Desulfurization?

Coal is unquestionably the fossil fuel of the future for generation of electrical energy, but its utilization gives rise to a number of ecological problems. These range from acid mine drainage, subsidence and destruction of land surface as a consequence of mining operations, to ash waste problems and air pollution from particulate and sulfur dioxide emissions resulting from combustion in utility boilers. It is estimated that worldwide sulfur dioxide emissions from coal combustion in 1965 amounted to 102×10^6 tons, a two-fold increase since 1940 (1).

The following sections describe the sulfur content and sulfur type distribution of coals on both a worldwide and U.S. basis, as well as the necessity for sulfur removal from coal, and the potential advantages of chemical desulfurization.

I. WORLD COAL RESERVES AND SULFUR CONTENT

A. Coal Reserves and Type

Table 1-1 lists the remaining "producible" coal reserves of the world. As indicated in Table 1-1, three super-powers (the United States, the People's Republic of China and the USSR) have 80% of the remaining coal reserves of the world; underdeveloped regions, such as Africa and South and Central America, are underendowed with this form of fuel.

TABLE 1-1
Remaining Coal Reserves of the World by Region and Principal
Coal-Producing Countries (3)

Region and country	Producible coal 10^9 metric tons	Percent of regional total	Percent of world total
Asia			
U.S.S.R.	600	52.3	25.8
China	506	44.1	21.8
India	32	2.8	1.4
Japan	5	0.4	0.2
Others	4	0.4	0.2
Total	1,147	100.0	49.4
North America			
United States	753	94.4	32.5
Canada	43	5.4	1.8
Mexico	2	0.2	0.1
Total	798	100.0	34.4
Europe			
Germany	143	47.5	6.2
United Kingdom	85	28.2	3.7
Poland	40	13.3	1.7
Czechoslovakia	10	3.3	0.4
France	6	2.0	0.3
Belgium	3	1.0	0.1
Netherlands	2	0.7	0.1
Others	12	4.0	0.5
Total	301	100.0	13.0
Africa	35	–	1.5
Australia	29	–	1.3
South and Central America	10	–	0.4
World Total	2,320		100.0

It should be pointed out that lower estimates of coal reserves are obtained by limiting the estimation to coal which can be economically mined using current technology. For example, U.S. coal reserves are estimated at only 150×10^9 tons of "hypothetical" reserves (2). Thus, the "producible" type estimates may be on the high side.

Current production is also dominated by the super-powers (Table 1-2), with West and East Europe completing the list of major producers. Combined production from these countries accounted for nearly 60% of the total world output in 1973.

TABLE 1-2
Major Coal Producing Countries in 1973 (4)

Region and country	Production (10^6 short tons)
Asia	
USSR	736
China (Mainland)	450
North America	
U.S.	592
Europe	
Germany (West)	231
Germany (East)	272
United Kingdom	140
Poland	216

The potential future importance of coal as a fossil fuel is shown by considering the life of producible coal reserves at a 0% production growth rate. The U.S.A. has approximately 1300 years of reserves, the People's Republic of China 1100 years, the USSR 800 years, the two Germanys 240 years, the United Kingdom 600 years and Poland 190 years of reserve.

B. Sulfur Distribution

Although many thousands of coal samples have been taken from operating coal mines, core samples, outcroppings, etc. and are reported in the literature, the sulfur content of the coal reserves of the world is not known with great accuracy. To obtain this data, it would be necessary to obtain an additional order of magnitude of coal samples for analysis, using a statistically valid sampling procedure, or more rigorously (and facetiously) to dig up, sample and analyze all the coal reserves of the world. Nevertheless, it is instructive to consider available data for both total sulfur content and the two major sulfur forms in coal (pyritic and organic sulfur) obtained for some selected coals from each major producing region (Table 1-3). The coals shown in the table are not meant as representative of any particular region or country, but are representative of worldwide distributions of sulfur.

It can be seen that the total sulfur content of these coal samples varies from 0.38% to a high of 5.32%. This is essentially the range of sulfur content which is normally found among coal samples on either a worldwide or regional basis. The pyritic sulfur content of these selected coals varies from a low of 0.09% to a high of 3.97%, while the organic sulfur content varies from a low of

TABLE 1-3
Sulfur Forms in Selected Bituminous Coal — Worldwide

Region and country	Location or mine	Sulfur, percent w/w[a]			Ratio Pyritic to Organic Sulfur
		Total	Pyritic	Organic	
Asia					
USSR	Shakhtersky (5)	0.38	0.09	0.29	0.031
China (Mainland)	Taitung (5)	1.19	0.87	0.32	2.7
India	Tipong (5)	3.63	1.59	2.04	0.78
Japan	Miike (5)	2.61	0.81	1.80	0.45
Malaysia	Sarawak (5)	5.32	3.97	1.35	2.9
North America					
U.S.	Eagle No. 2 (6)	4.29	2.68	1.61	1.7
Canada	Fernie (5)	0.60	0.03	0.57	0.053
Europe					
Germany	――― (7)	1.78	0.92	0.76	1.2
United Kingdom	Derbyshire (8)	2.61	1.55	0.87	1.8
Poland	――― (5)	0.81	0.30	0.51	0.59
Africa					
So. Africa	Transvaal (9)	1.39	0.59	0.70	0.84
Australia	Lower Newcastle (10)	0.94	0.15	0.79	0.19
South America					
Brazil	Santa Caterina (11)	1.32	0.80	0.53	1.5

[a]Moisture-free basis, pyrite + sulfate reported as pyrite

0.29% to a high of 2.04%. Generally speaking, organic sulfur levels much greater than 2% or much less than 0.3% are almost never encountered, and pyritic sulfur levels greater than 4% are also uncommon. However, the pyrite content of a few coals can approach zero when there is both little inherent pyrite and when careful mining operations (such as manual labor) prevent the mining of pyrite-containing formations adjacent to the coal seam. The ratio of pyritic to organic sulfur can vary over 2-3 orders of magnitude.

The sulfur content and sulfur forms distribution of U.S. coals have been more extensively reported than those of other countries (12, 13, 14, 15). Still, no set of data is available which fully and statistically describes both the sulfur content and sulfur distribution for U.S. coals.

Some sulfur analyses of run-of-mine coal recently obtained from major operating mines representative of the three most important U.S. coal regions are shown in Figure 1-1. Emphasis is placed on Appalachian coal samples, since the Appalachian region supplies more than 60% of current U.S. coal production. It can be seen that the overall sulfur content of western coals is low, generally below 1.0%, and the major sulfur form is organic. Run-of-mine Interior Basin coals generally have a sulfur content of about 4.0%, with about 30-50% of the sulfur being organic. Appalachian coals cover a larger range of sulfur content and ratio of sulfur forms, but tend to have organic sulfur content lower than either the pyritic sulfur or the organic sulfur levels of Interior Basin coals. Thus, the geographical trend in sulfur content from east to west is high sulfur to low sulfur and a preponderance of pyritic sulfur to a preponderance of organic sulfur.

As noted above, a majority of current U.S. coal production comes from the Appalachian region, i.e., the states of Virginia, Tennessee, Pennsylvania, Kentucky (east) and West Virginia. The sulfur content of Pennsylvania coal is typical of the region (Figure 1-2). The organic sulfur levels tend to center around 0.5-1.0% and pyritic sulfur level varies from nearly zero to more than 3.0%, giving rise to total sulfur content mainly at the 1.0-4.0% level. Here it becomes apparent that the removal of one type of sulfur alone, the pyritic sulfur, can reduce a large number of these coal samples to less than 1.0% total sulfur, a value consistent with many air pollution control standards.

II. WHY REMOVE SULFUR?

Sulfur, in the form of its element or combined with other elements, is a nutrient for both plant and animal life. However, in recycling sulfur (as in other nutrients) back to nature, ecological soundness requires that there must be no excess at any given point in the cycle. In general, the fate of sulfur dioxide emissions involves photo-oxidation in the atmosphere to form sulfur trioxide which, under humidifying conditions, becomes sulfuric acid or sulfate aerosol. Residual sulfur dioxide and sulfuric acid or sulfates are scavenged from the atmosphere by vegetation, eventually being discharged into the sea by the world's rivers, along with sulfur accumulated from weathering rocks and sulfur applied as fertilizer. Some of the sulfate is directly deposited into the ocean via rain or dust. When an excess occurs, the atmosphere-to-land portion of the sulfur cycle is unsound.

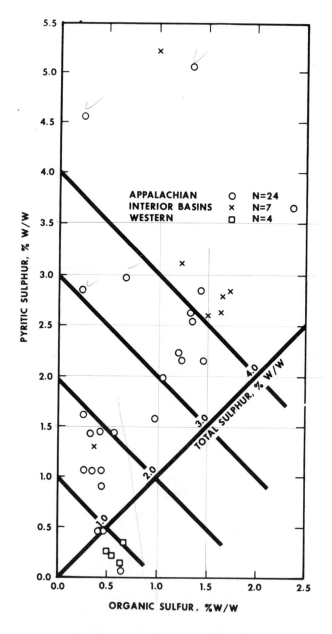

Figure 1-1.   Distribution of sulfur forms (dry moisture free basis)
in run-of-mine U.S. coals (6)

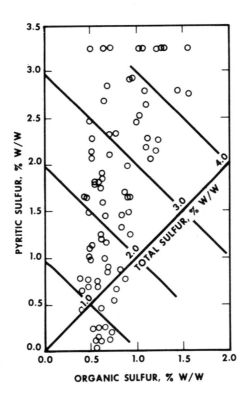

Figure 1-2. Distribution of sulfur forms (dry moisture free basis) in run-of-mine Pennsylvania utility coal (12)

A. Air Pollution Control

Worldwide sulfur air emissions are summarized in Table 1-4. Coal combustion comprises about 67% (on a sulfur tonnage basis) of man-made sulfur pollution and about 23% of all sulfur air emissions. At first glance, coal combustion would thus appear to be only a small part of the sulfur compound air emission problem. However, coal combustion sources, i.e., electric utilities, industrial boilers and commercial boilers, are heavily concentrated in and around the industrial cities of the world which also, of course, have the highest population density. By contrast, the natural marine and terrestrial emissions are dispersed from the outset over a wide geophysical area. This is with the exception of volcanic activity which is intermittent and a minor source of hydrogen sulfide and sulfur oxide emissions, on an average basis.

TABLE 1-4
Estimated Worldwide Sulfur Compound Air Emissions, 1965 (1)

Compound	Source	Emissions as sulfur (tons/yr x 10^6)
SO_2	Coal combustion	51
	Petroleum refining	3
	Petroleum combustion	11
	Smelting of ores	8
H_2S	Industrial emissions	3
	Marine emissions	30
	Terrestrial emissions	70
SO_4 aerosols	Marine emissions	44
	Total air emissions	220

In the United States, coal combustion accounts for roughly 60% of sulfur oxide air emissions (Table 1-5). This amounted to 14.0 x 10^6 tons in 1966 and 16.5 x 10^6 tons (16) of sulfur dioxide in 1972.

To contend with this pollution problem, the Federal Government has established a performance standard for all new coal-fired plants greater than 25 MW of 1.2 lbs $SO_2/10^6$ Btu of coal heat input (17). For most Eastern coals this corresponds to a maximum allowable coal sulfur content of approximately 0.8% and about 0.6% for Western coals. As indicated in Figures 1-1 and 1-2, very little of the current run-of-mine utility coal production in the United States meets this sulfur content restriction.

TABLE 1-5
Estimated U.S. Sulfur Dioxide Air Emissions, 1966 (18)

Source	Percent of Emissions
Coal Combustion	58.2
Petroleum Combustion	19.6
Petroleum Refining	5.5
Smelting of Ores	12.2
Coke Processing	1.8
Sulfuric Acid Manufacture	1.9
Miscellaneous	0.4

B. Coking Coal

The charge to the blast furnace consists mainly of three materials: iron, limestone and coke. The limestone converts to lime during the heating process and becomes a part of the slag, along with the ash and adsorbed sulfur from the coke.

The economics of the production of iron is strongly dependent on the sulfur content of coke utilized in the blast furnace. Because pig iron cannot contain more than a small amount of sulfur, the sulfur content of coal-derived coke can be no more than roughly 1.0%. The sulfur content of coke in turn depends on the amount of sulfur present in coal fed to the coke oven. Product coke contains about 75% of the sulfur level of the feed coal.

When coke fed to the blast furnace contains excess sulfur, relatively large amounts of additives such as lime and more coke must be added to the blast furnace to insure that the sulfur content of the pig iron is maintained at a low level. This in turn decreases the throughput of the blast furnace and has an economic impact. The relationship of blast furnace throughput to sulfur content of the coke can be seen in Figure 1-3 (19) where the amount of coke which must be used per ton of product iron increases as the sulfur content of coke rises and the amount of slag produced per ton of pig iron also increases. Both of these have the effect of decreasing the rate of production of iron.

Desulfurization of coking coal prior to coke production can be economically desirable. Most coking coal is physically cleaned prior to use in order to reduce the amount of mineral component (often termed "ash") present in the coal, and at the same time to remove some pyritic sulfur which is associated with the mineral component (physical cleaning of coal is further discussed in the next section). Physical cleaning data (obtained by a float-sink test procedure) for three coking coals is shown in Figure 1-4. The organic sulfur level for all three coals is below 1%, so they are all theoretically washable to below 1% total sulfur. However, the Corona seam and South American coals cannot be washed to a sulfur content below 1.5%, even at a vanishingly small yield. This is due to limited liberation of pyrite in these coals by the grinding procedure. This is in contrast to the Lower Kittanning seam coal, which can be washed to below 1.0% sulfur with a reasonable coking coal yield of 25-30%.

However, if a process were available for chemical removal of just the pyritic sulfur from the Corona seam or South American coal, their sulfur content could be reduced to below 1% and thus, these resources could be made available for the production of pig iron and steel. For example, a 50% yield of coking coal for either of these coals could be obtained by chemical desulfurization of the 1.9 float fraction, while a 25-30% yield of coking coal could be obtained by desulfurizing the 1.4 float material. The 1.4 float material from each of these two coals contains lower ash and so would be the more likely candidate for production of desulfurized coking coal.

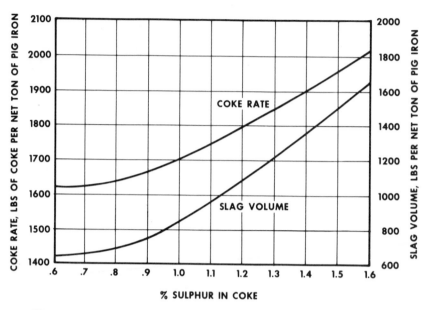

Figure 1-3. Effect of sulfur in coke on coke rate and slag volume (19)

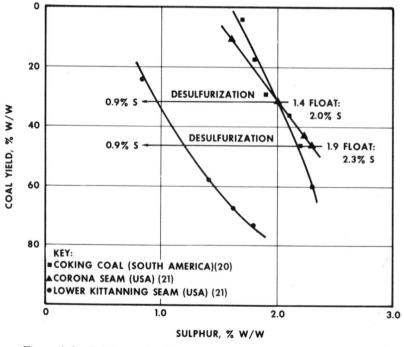

Figure 1-4. Coking coal sulfur reduction by cleaning (1-1/2 in x0) (20, 21)

III. WHY CHEMICAL DESULFURIZATION?

Pollution of the atmosphere arising from coal combustion can be combatted in a number of ways including: flue gas desulfurization by scrubbing, the building of tall stacks to disperse the pollutant over a wider area, conversion of coal to low sulfur (and low ash) fuel oil, and coal cleaning. The technology for these approaches is in varying stages of development, from laboratory and pilot scale in the case of liquefaction, and demonstration plant development for flue gas scrubbing, to a fully developed and mature industry in the case of physical cleaning of coal. Chemical desulfurization is most comparable to this latter technology, which is discussed below.

Physical cleaning of coal to remove ash along with a part of the pyrite content has been commercial practice for a number of years. Because today's continuous and non-selective mining machines result in a substantial impurity content in coal, the majority of utility fuel is now cleaned prior to combustion. Coal cleaning (or washing) is the separation of raw coal into ash-rich and ash-lean fractions by virtue of the relatively high density and hardness of ash in relation to the hydrocarbon coal matrix. This is done in commercial practice through first crushing to liberate some pyrite, then passing the coal through screens, jigs, dense-medium processes, tables, etc. The behavior of coal in mechanical cleaning plants is approximated in the laboratory through float-sink testing of crushed coal in liquids of various density.

An extensive survey of the sulfur reduction potential of the coals of the United States was performed by Deurbrouck at the U.S. Bureau of Mines (12). Coals were selected from all major coal producing regions of the United States. The average coal contained 2.05% pyritic sulfur and 3.23% total sulfur; the raw coal content averaged 63% pyritic sulfur and 37% organic sulfur. Only about 7% of the as-mined coals were in the 0.8% sulfur range needed to meet the Standards for New Stationary Sources (17).

Each sample was tested at several media densities and coal top-sizes. The results are summarized in Figure 1-5. Current coal cleaning technology is best represented by the float-sink curves (Curves c of Figure 1-5) at 1-1/2 inch top-size, since it is not considered practical to clean 14 mesh top-size as the primary input to a coal cleaning plant. It can be seen that about 15% of the mines could be washed to a sulfur level of 0.8% if a 60% coal yield was allowable. This value is increased to about 20% of the mines for 14 mesh x 0 coal. If coal were cleaned to a 60% coal yield, however, this would pose a serious solid waste disposal problem as well as a continual sulfur oxide pollution problem, as waste piles slowly give off sulfur dioxide while smoldering.

12

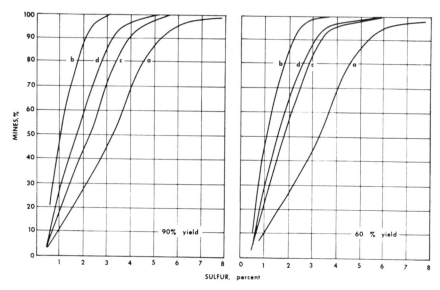

Figure 1-5.   Washability Summary of 322 U.S. Coals (12). Raw coal total sulfur
content curve (a), the raw coal organic sulfur content curve (b), total
sulfur content curves at two coal yields: raw coal was crushed to
1-1/2-in top-size (c) and 14-mesh top-size (d).

If a method were available for the chemical removal of pyritic sulfur from coal, the "washability" curves reduce to the organic sulfur level shown in Curves b of Figure 1-5, resulting in roughly 35% of the mines being reduced to a sulfur content no greater than 0.8% and this would be at nearly 100% coal yield. The loss (if any) is dependent on the type of chemical removal and its selectivity. Additional removal of about half of the organic sulfur would cause nearly all coals to fall below 0.8% total sulfur.

It might be hoped that additional size reduction would result in further liberation of pyrite and, therefore, increased reduction in the pyritic sulfur level of coal. The effect of coal top-size reduction on the washability of Midwest region coals is shown in Figure 1-6, which indicates that the liberation of pyrite increases as coal top-size is reduced. However, float-sink data on Midwestern and Western coals (shown in Table 1-6) indicate that while further pyrite reduction is usually obtained on coal size reduction, substantial pyrite is retained in the float portion of the coal matrix even at very fine coal top sizes.

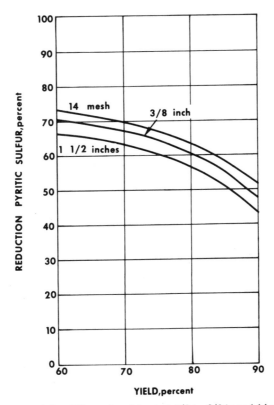

Figure 1-6. Effect of crushing to 1-1/2-in, 3/8-in and 14-mesh top-sizes on liberation of pyritic sulfur, Midwest regional Coals (12)

TABLE 1-6

Pyritic Sulfur Reduction by Cleaning – Effect of Coal Top-Size (21)

Coal Seam	Raw Coal	Pyritic sulfur content, percent w/w Cleaned at various top-sizes[a]		
		3/8 in	14 mesh	p.c. grind[b]
Bevier	3.5	1.5	1.3	0.8
Des Moines No. 1	2.6	1.3	1.2	1.3
Lower Cherokee	3.5	1.7	1.5	0.8
Fort Scott	4.7	2.4	2.2	1.4
Tebo	3.1	1.0	0.7	0.4
Lower Spadra	3.3	2.4	2.1	0.5
Charleston	2.9	2.5	2.3	1.2

[a] 1.60 float

[b] 70% −200 mesh

REFERENCES

1. S.F. Singer, ed., Global Effects of Environmental Pollution, Springer-Verlag, Inc., New York, 1970, p. 54.

2. John E. Emery, ed., Coal Age, 79: 73 (1975).

3. P. Averitt, Coal Reserves of the United States: A Progress Report, U.S. Geol. Survey Bull., 1136, January 1960.

4. G.F. Nielsen, ed., Keystone Coal Industry Manual, McGraw-Hill Mining Publications, New York, 1975, pp. 158-9.

5. Private communication, 1974.

6. J.W. Hamersma and M.L. Kraft, "Applicability of the Meyers Process for Chemical Desulfurization of Coal," EPA-650/2-74-025a, U.S. Environmental Protection Agency, Washington, D.C., 1975.

7. F. Foerster and W. Geisler, Ztch. Angew. Chem., 36: 193, (1922).

8. T.G. Woolhouse, Fuel in Science and Practice, 4: 456, (1925).

9. H.F. Yancy and M.R. Geer in Coal Preparation, (J.W. Leonard and D.R. Mitchess, eds.), Am. Inst. Min. Met. and Petrol. Engs. Inc., New York, 1968, pp. 1-47.

10. "Eastern Australian Coals," Joint Coal Board and Queensland Coal Board, 1974, p. 27.

11. S.C. Trindade, Metalurgia, 31: 385, (1975).

12. A.W. Deurbrouck, "Sulfur Reduction Potential of the Coals of the United States," Report of Investigations 7633, Bureau of Mines, U.S. Department of the Interior, Washington, D.C., 1972.

13. H.J. Gluskoter and J.A. Simon, American Chemical Society, Division of Fuel Chemistry, Preprints, 8: 159, (1964).

14. "The Reserve Bases of U.S. Coals by Sulfur Content," U.S. Bureau of Mines Information Circular 8680, U.S. Department of Interior, Washington, D.C., 1966.

15. "Forms of Sulfur in U.S. Coals," U.S. Bureau of Mines Information Circular 8301, U.S. Department of the Interior, Washington, D.C., 1966.

16. "National Emissions Report," EPA 450/2-74-012, U.S. Environmental Protection Agency, Washington, D.C., June 1974.

17. U.S. Environmental Protection Agency, Federal Register, 36: 24876 (1973).

18. "Air Quality Criteria for Sulfur Oxides," National Air Pollution Control Administration Publication No. AP-50, U.S. Department of Health, Education and Welfare, Washington, D.C., January 1969.

19. T. H. Kennedy and A. W. Thornton, Yearbook, American Iron and Steel Institute, New York, 1949, p. 222.

20. Personal communication, R. A. Meyers.

21. "An Evaluation of Coal Cleaning Process and Techniques for Removing Pyritic Sulfur from Fine Coal," National Air Pollution Control Administration, Bituminous Coal Research Inc., PH86-76-139, Washington, D.C., 1971.

CHAPTER 2

COAL STRUCTURE AND CHEMICAL DESULFURIZATION

I. Chemical Structure
    A. The Organic Coal Matrix
    B. Inorganic Coal Structure
II. Physical Structure of Coal

Coal is a highly heterogeneous solid originating from plant substance. It contains, in varying amounts, essentially all elements of the periodic table combined into nearly all of the minerals normally encountered in the earth's crust. The organic matrix comprises most of the coal weight and consists mainly of carbon, with smaller amounts of hydrogen, oxygen, nitrogen and sulfur. Coal can be considered a rock structure containing both macroscopic and microscopic petrographic features. It has also been viewed as an organic chemical substance containing the classical organic functional groups, e.g., mainly carbonyl and hydroxyl, aromatic and heterocyclic ring units and aliphatic bridges. From another standpoint, coal is a solid colloid which has a large volume porosity and can adsorb gases and vapors as well as liquids. Furthermore, the organic coal matrix may be characterized as a crosslinked polymer (formed from the cellulosic polymer present in plant material) which, in the absence of degradation, is essentially insoluble and nonvolatile.

Some excellent reviews of coal structural features are available (1, 2, 3) and the reader is directed to these for a thorough understanding of the subject. The following sections will concentrate on the structure of the sulfur-containing portions of coal with reference to the overall system only as necessary for an understanding of the sulfur microcosm.

I. CHEMICAL STRUCTURE

A. The Organic Coal Matrix

Given (4) has proposed an admittedly arbitrary model (Figure 2-1) for the organic coal matrix involving methylene bridges of the 9, 10-dihydroanthracene type, aromatic structures including benzopyridine, benzoquinone and benztropolone as major constituents, and nonaromatic units such as cyclohexanone, cyclohexane and the like. The major feature of this model involves the bonding of aromatic nuclei by two methylene linkages.

This structure was suggested on the basis of spectroscopic and carbon/hydrogen ratio data, as well as a knowledge of the chemical reactions of coal with bromine and oxygen. A complete determination of the actual structure formula for coal is an elusive goal, since coal structure must be highly variable among various coals and may include all known organic structures, as well as some yet to be defined.

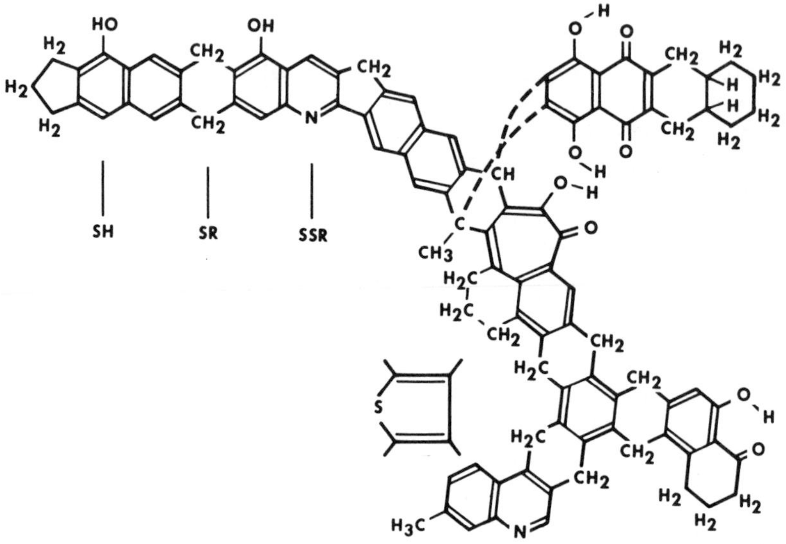

Figure 2-1. Model of Organic Coal Matrix (4)

In a subsequent paper (5) Given summarized the available knowledge of the structure of the organic sulfur containing functions present in the coal organic matrix. Mercaptan, sulfide, disulfide and thiophene (see Figure 2-1) were suggested as the major organic sulfur-containing functional groups. One additional sulfur-containing ring, thiopyrone, was also proposed.

Some idea of the potential number of organic sulfur compounds which may be present in coal can be gained from considering those identified in crude oil (Table 2-1). Although crude oil is of animal origin, its structural complexity is no more diverse than coal.

TABLE 2-1

Sulfur Compounds* Identified by Bureau of Mines – API
Research Project 48 on Four Crude Oils (6)

Class	Wasson, Texas	Wilmington, California	Agha Jari, Iran	Deep River, Michigan	Total
Thiols					
Alkyl	39	–	6		45
Cyclic	7	–			7
Aromatic	1	–			1
Sulfides					
Alkyl	38	–	5		43
Alkyl-Cycloalkyl	5				5
Cyclic	21	14			35
Disulfides	1			3	4
Thiophenes	–	11			11
Benzothiophenes	13				13
Dibenzothiophenes	2				2
Total	127	25	11	3	166

*Includes both "tentative" and "firm" identification.

B. Inorganic Coal Structure

The major minerals present in U.S. bituminous coals are listed in Table 2-2. It is very difficult to totally separate the minerals from the coal for analysis since the mineral component of coal is intimately associated with the organic coal matrix. The listing in Table 2-2 is based on optical microscopy, spectroscopy and analysis of the ash content obtained after combustion of the organic coal matrix. Coal ash composition ranges for oxides of the above minerals are shown in Table 2-3.

TABLE 2-2

Minerals Associated with Bituminous Coals (7)

Group	Species	Formula
Shale	Muscovite Hydromuscobite Illite Bravaisite Montmorillonite	$(K, Na, H_3O \ Ca)_2 \ (Al, Mg, Fe, Ti)_4 \ (Al, Si)_8 O_{20} \ (OH, F)_4$
Kalin	Kaolinite Livesite Metahalloysite	$Al_2(Si_2O_5) \ (OH)_4$
Sulfide	Pyrite Marcasite	FeS_2
Carbonate	Ankerite Ankeritic calcite Ankeritic dolomite Ankeritic chalybite	$(Ca, Mg, Fe, Mn) \ CO_3$
Chloride	Sylvine Halite	KCl NaCl

TABLE 2-3

Coal Ash Composition: Oxides of Minerals
Associated with Bituminous Coals (8)

Constituent Oxide	Percent
Silica (SiO_2)	20 – 60
Aluminum oxide (Al_2O_3)	10 – 35
Ferric oxide (Fe_2O_3)	5 – 35
Calcium oxide (CaO)	1 – 20
Magnesium oxide (MgO)	0.3 – 4
Titanium oxide (TiO_2)	0.5 – 2.5
Alkalies ($Na_2O + K_2O$)	1 – 4
Sulfur trioxide (SO_3)	0.1 – 12

The trace element composition of coal is highly variable, depending on the method of mining, origin, etc.. Trace element analyses recently obtained on some U.S. coals are shown in Table 2-4. Many of these elements are associated with both the inorganic and organic portions of the coal matrix.

As previously shown in Table 2-2, the sulfur content of the inorganic portion of coal occurs mainly in the minerals pyrite and marcasite. A small amount of inorganic sulfur is also present as sulfate minerals such as melanterite ($FeSO_4 \cdot 7H_2O$) and jarosite $[(Na, K)Fe_3(SO_4)_2(OH_6)]$, as well as calcium sulfate (gypsum). In freshly-mined coal, the total sulfur content, as sulfate, rarely exceeds 0.1 percent by weight of coal, but in weathered coal, it may rise to 0.2 percent or more. The minerals pyrite and marcasite have the same chemical composition (FeS_2) but differ in crystal form; pyrite is isometric (cubic), while marcasite is orthorhombic. Of the two, pyrite is the most commonly found form in coal, with marcasite occuring in lesser amounts. Marcasite, slightly less stable than pyrite and more reactive chemically, is converted to the pyrite form on heating. Throughout the rest of this discussion both forms will be referred to by the common term, pyrite.

II. PHYSICAL STRUCTURE OF COAL

As previously noted, coal has an extensive pore structure. Fresh coal, although it may appear dry and dusty when crushed, contains a large amount of absorbed water. The moisture content ranges from 1 to 5 percent in bituminous coal, 20 percent in subbituminous to nearly 45 percent by weight in lignite. Brooks (10) reports that high organic sulfur content coals tend to have a lower moisture content than low organic sulfur coals of the same rank. This was explained by the fact that since hydrogen bonding (by water) to sulfur groups is very weak or nonexistent, as evidenced by the lower water solubility of simple sulfur compounds compared to their oxygen analogs, high organic sulfur coals could be relatively more hydrophobic than low organic sulfur coals. This fact becomes important when considering the potential for chemical removal of organic sulfur from coal (see Chapter 3) which requires penetration of the coal matrix with liquids in order to cause chemical reaction with the organic sulfur. Brooks' postulate would indicate that high organic sulfur coal would be more permeable to organic solvents(hydrophobic), than to aqueous chemical solutions.

The pyrite content of coal is found in both macroscopic and microscopic dispersions (8). Macroscopic pyrite occurs predominantly in four forms: 1) veins, thin and film-like, or up to several inches wide; 2) lenses, generally flattened and elongate in cross-section, ranging from fractions of an inch to several inches; 3) nodules or balls, varying in size from inches to several feet in diameter; and 4) pyritized plant tissue appearing with carbonate minerals in a "coal ball."

The size distribution of pyrite (see Figure 2-2) is believed to tend toward small particle size for low pyritic sulfur coal and towards large particle size for high pyritic sulfur coal. This point is significant in the chemical removal of pyritic sulfur where the reaction rate may depend on the particle size of the

TABLE 2-4

A Trace Element Composition of Some U.S. Coal Mines (PPM) (9)

Element	Appalachian coal basin														Eastern Interior basin				Western coals	
	Muskingum	Mathies	Robinson	Powhattan	Delmont	Marion	Lucas	Bird No. 3	Meigs	Egypt Valley No.21	Jane	Fox	Warwick	Humphrey No. 7	Ken	Eagle No. 2	Orient No. 6	Camp Nos. 1 & 2	Belle Ayr	Colstrip
Ag	2.3	1.8	1.6	0.8	2.6	1.5	2.0	2.9	0.6	4	2	<0.1	4	0.5	1.4	<0.1	<0.1	8	<0.1	<0.1
As	2.0	6.1	5.9	4.3	40	98	74	16	2.6	22	29	24	13	9	6.5	6.6	15.2	5.7	0.4	<0.01
B	54	54	60	62	18	10	20	30	115	34	27	16	20	26	6.0	30	43	272	11	35
Be	2.0	2.7	0.6	3.3	4.2	2.2	3.8	3.6	1.4	0.7	0.8	2.0	1.0	0.4	2.0	0.5	6	1.5	<0.5	0.5
Cd	1.6	0.8	1.8	1.2	1.8	1.5	1.4	1.4	0.8	<0.5	<0.5	<0.5	<0.5	<0.5	1.7	0.5	0.7	0.8	0.6	<0.5
Cr	110	110	100	141	144	76	52	149	100	55	55	94	81	26	76	126	74	122	<0.5	<0.5
Cu	15	29	10	25	20	38	13	26	23	26	35	25	24	16	16	18	36	17	27	8
F	117	210	100	282	131	155	65	105	222	168	122	94	251	78	124	151	105	215	48	29
Hg	0.09	0.09	0.14	0.07	<0.02	0.06	<0.2	0.10	0.05	0.31	0.11	0.07	0.14	0.06	<0.2	0.16	0.12	0.16	0.22	0.20
Li	55	64	12	52	24	76	8	54	22	26	38	4	76	13	9	4	23	10	<0.3	4
Mn	25	66	42	57	94	25	15	45	44	41	46	24	31	31	60	86	57	98	48	87
Ni	29	34	26	37	68	23	35	36	41	41	33	147	44	17	30	136	53	27	61	31
Pb	12	19	12	20	31	15	18	23	12	15	25	5	16	7	16	29	0.5	25	3	3
Sb	<5	<5	19	<5	16	<5	<5	<5	9	<5	<5	<5	<5	<5	24	<5	<5	<5	<5	<5
Se	59	74	49	54	25		8		63	<5	<5	17	<5	<5	15	<5	<5	<5	5	<5
Sn	15	12	8	<5	20	<5	10	15	15	<5	<5	<5	<5	<5	12	<5	<5	<5	<5	<5
V	33	60	28	60	40	54	12	60	50	102	147	94	78	77	35	64	69	105	1	103
Zn	30	41	30	40	76	34	50	80	38	31	34	105	55	18	40	215	25	97	49	9

pyrite rather than on the particle size of the coal itself; that is, the desulfurization reaction rate may be controlled by both the rate of diffusion of the chemical into (and out of) the coal matrix and the surface area of the pyrite particles. Thus, coals containing small amounts of pyritic sulfur as a result of previous coal cleaning or inherent nature may react faster than high pyritic sulfur coals.

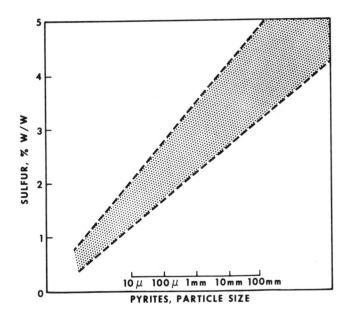

Figure 2-2. Typical Distribution of Pyrite in Coal (11)

Coal mine pyrite is quite porous relative to mineral pyrite. The surface area of a specimen of coal mine pyrite from a U.S. mine was found to be about 40,000 cm^2/g at an average diameter of 60 microns and 1000 to 2000 cm^2/g at a diameter of 700 microns (12). This is about an order of magnitude greater surface area than that found for a museum grade pyrite specimen. The specific gravity of the coal pyrite was 3.2 to 3.4, compared to 5.0 for pure mineral pyrite.

REFERENCES

1. D. W. Van Krevelen, Coal, Elsevior Publishing Co., New York, (1961).

2. H. H. Lowry, ed., Chemistry of Coal Utilization, Vol. I, John Wiley and Sons, New York, (1963).

3. H. H. Lowry, ed., Chemistry of Coal Utilization, Supp. Vol., John Wiley and Sons, Inc., New York, 1963.

4. P. H. Given, Fuel, 39: 147 (1960).

5. P. H. Given and W. F. Wyss, Brit. Coal Utilization Research Association Monthly Bulletin, 25: 165 (1961).

6. Bureau of Mines – API Research Project 48.

7. J. B. Nelson, BCURA Bull., 17: 43 (1953).

8. J. W. Leonard and Dr. R. Mitchell, eds., Coal Preparation, Am. Inst. of Min., Met. and Petrol. Engs., New York, 1968, p. 1-37.

9. J. W. Hamersma and M. L. Kraft, "Applicability of the Meyers Process for Chemical Desulfurization of Coal: Survey of Thirty-Five Coals," EPA-650/2-74-025a, U.S. Government Printing Office, Washington, D.C., 1975.

10. J. D. Brooks, J. Inst. Fuel, 29: 82 (1956).

11. W. Chapman and D. Rhys Jones, J. Inst. Fuel, 28: 102 (1955).

12. C. S. Clark, J. San. Div. Am. Civil Engrs., 92: 127 (1966).

CHAPTER 3

CHEMISTRY OF DESULFURIZATION REACTIONS

I. Chemical Reactions of Pyrite
    A. Classification of Pyrite Reactions
    B. Oxidation of Pyrite
        1. Electrochemical Potential
        2. Chemical Reactions
            a. General Reactions
            b. Reactions with Ferric Salts
            c. Reactions with Oxygen
    C. Reduction of Pyrite

II. Desulfurization of Organic Sulfur Compounds
    A. Classification of Desulfurization Methods
        1. Solvent Partition
        2. Thermal Decomposition
        3. Acid-Base Neutralization
        4. Reduction
        5. Oxidation
        6. Nucleophilic Displacement
    B. Desulfurization Reactions
        1. Solvent Partition
        2. Thermal Decomposition
        3. Acid-Base Neutralization
        4. Reduction
        5. Oxidation
        6. Nucleophilic Displacement

As indicated in Chapters 1 and 2, sulfur occurs in two major forms (pyrite and organic) in coal which are separately located in the inorganic and organic portions of coal, respectively. This division of the coal matrix into inorganic and organic portions plays a significant role in the chemistry of coal and is particularly significant in the design of methodology for the penetration, contact and chemical reaction of the various sulfur functional groups. Conceivably, a chemical reagent could be dissolved in a hydrophobic solvent such that the reagent comes into intimate contact only with the organic carbon matrix of the coal, while another reagent, dissolved in water, would come into more intimate contact with the inorganic portion of the coal matrix. This could then be a means of obtaining part of the selectivity required for an effective economic desulfurization process. However, because of the wide reactivity of a given chemical reagent, it is not possible to obtain a high degree of chemical selectivity between the organic and inorganic portions of the coal matrix in many cases. The following sections discuss a variety of chemical reactions which can effect desulfurization of inorganic or organic funnctional groups present in coal.

I. CHEMICAL REACTIONS OF PYRITE

The chemical reactions of iron pyrite are numerous. The following discussion places special emphasis on those reactions which have been observed or are theoretically capable of proceeding at temperatures below the 400°C decomposition point of coal.

A. Classification of Pyrite Reactions

A solvent which will dissolve significant amounts of iron pyrite without decomposition is yet to be discovered. Therefore, removal of pyritic sulfur from coal requires a chemical transformation such as:

a)   A displacement reaction (Eq. 1)

$$:N: \quad Fe\text{-}S_2 \rightarrow :N: Fe + S_2^{-2} \tag{1}$$

b)   An acid base neutralization, which may produce hydrogen sulfide, an iron salt, (Eq. 2)

$$4H^+ + FeS_2 \rightarrow 2H_2S + Fe^{+2} \tag{2}$$

c)   An oxidation reaction which converts the persulfide ion to a soluble, volatile or otherwise removable form, e.g., (Eq. 3) or

$$[O] + FeS_2 \rightarrow Fe^{+2} + 2S + [O^{-2}] \tag{3}$$

d)   A reduction reaction which would remove pyritic sulfur as hydrogen sulfide (Eq. 4, 5).

$$Fe\,S_2 + H_2 \rightarrow Fe\,S + H_2S \tag{4}$$

$$Fe\,S + H_2 \rightarrow Fe + H_2S \tag{5}$$

The reactions of iron pyrite reported in the literature are almost exclusively oxidation and reduction reactions. No displacement or neutralization reactions are reported. The chemistry of oxidation and reduction of iron pyrite is described in the following two sections.

B. Oxidation of Iron Pyrite

1. ELECTROCHEMICAL POTENTIAL

The sulfur content of iron pyrite can be oxidized to form elemental sulfur (Eq. 6) or sulfate (Eq. 7).

$$Fe\ S_2 \longrightarrow Fe^{+2} + 2S + 2e^- \tag{6}$$

$$Fe\ S_2 + 8\ H_2O \longrightarrow Fe^{+2} + 2\ SO_4^{-2} + 16\ H^+ + 14\ e^- \tag{7}$$

The standard reduction potentials for these half-reactions have been calculated by Biernat and Robins (1,2) at +0.421 and +0.362, respectively. These reduction potentials are dependent on temperature, pH and concentrations of sulfate and ferrous ion. The SO_4^{-2}/FeS_2 couple can tend toward the H^+/H_2 potential at higher pH. Because of the relatively low reduction potentials of these couples, it would appear that a number of aqueous oxidants (Table 3-1) would be capable of oxidizing the persulfide portion and, in most cases, the ferrous component in aqueous solution.

Oxidants ranging from metal ions (Fe^{+3}, Hg^{+2}, Ag^+) through strong acids (HNO_3 and $HClO_4$) as well as O_2, Cl_2, SO_2 and H_2O_2 have sufficient oxidation potential to react with iron pyrite. However, potential is a measure of free energy difference and bears no relation to the kinetics of the reactions which may be observed.

At low acidity, Sn^{+4} and Cu^{+2} may also have the potential to oxidize pyrite. An important distinction between the oxidants lies in their potential for oxidation of Fe^{+2} and/or the S_2^{-2} components of iron pyrite. For example, H_2SO_3 (sulfurous acid) has sufficient oxidation potential to oxidize the persulfide function to either sulfur or sulfate but has insufficient potential to oxidize Fe^{+2}. Thus, the products formed by oxidation of iron pyrite with SO_2 would be Fe^{+2} and either elemental sulfur or sulfate or both. The standard reduction potentials of H_2SO_3 and SO_4^{-2} indicate that, should SO_2 be used as oxidant, a disproportionation reaction can occur giving sulfate and sulfur as products. Indeed, the disproportionation of SO_2 occurs in aqueous solutions and is catalyzed by metal ions (5). Thus, any oxidant with the potential to oxidize to the SO_2 state can be expected to form sulfate as well. Oxidants with potentials greater than +0.771 can oxidize Fe^{+2} to Fe^{+3} but may not, depending on kinetics, while those with potentials more positive than 0.45 can oxidize sulfur formed to SO_2 and sulfate.

TABLE 3-1

Standard Reduction Potentials in Aqueous Acid Solutions

$H^+ + e^- = 1/2\ H_2$	+0.00	(3)
$S + 2H^+ + 2e^- = H_2S_{(aq)}$	+0.141	(3)
$Sn^{+4} + 2e^- = Sn^{+2}$	+0.15	(3)
$Cu^{+2} + e^- = Cu^+$	+0.167	(3)
$SO_4^{-2} + 4H^+ + 2e^- = H_2SO_3$	+0.20	(3)
$Fe^{+2} + 2SO_4^{-2} + 16H^+ + 14e^- = FeS_2 + 8H_2O$	+0.362	(2)
$Fe^{+2} + 2S + 2e^- = FeS_2$	+0.421	(2)
$H_2SO_3 + 4H^+ + 4e^- = S + 3H_2O$	+0.45	(3)
$Fe^{+3} + e^- = Fe^{+2}$	+0.771	(3)
$1/2\ Hg_2^{+2} + e^- = Hg$	+0.7986	(3)
$Ag^+ + e^- = Ag$	+0.7995	(3)
$NO_3^- + 3H^+ + 2e^- = HNO_2 + H_2O$	+0.94	(3)
$1/2\ O_2 + 2H^+ + 2e^- = H_2O$	+1.229	(3)
$1/2\ Cl_2 + e^- = Cl^-$	+1.34	(3)
$ClO_4^- + 8H^+ + 7e^- = 1/2\ Cl_2 + 4H_2O$	+1.3583	(3)
$H_2O_2 + 2H^+ + 2e^- = 2H_2O$	+1.77	(4)

2. CHEMICAL REACTIONS

An excellent summary of early pyrite chemistry is presented in Mellor's, "Inorganic Chemistry" (6). The following section presents some of the more pertinent reactions.

a. General Reactions

The reactions of the oxidants, hydrogen peroxide (Eqs. 8, 9) and nitric acid (Eqs. 10, 11),

$$2FeS_2 + 3H_2O_2 \rightarrow Fe_2O_3 + 2S + 3H_2O \tag{8}$$

$$S + 6H_2O_2 + 4H^+ \rightarrow SO_4^{-2} + 8H_2O \tag{9}$$

$$2FeS_2 + 10\,HNO_3 \rightarrow Fe_2(SO_4)_3 + H_2SO_4 + 10\,NO + 4\,H_2O \tag{10}$$

$$S + 3HNO_3 + H_2O \rightarrow SO_4^{-2} + 3NO_2^- + 5H^+ \tag{11}$$

sulfuric acid and sodium hypochlorite are described, reportedly giving varying mixtures of Fe^{+2}, Fe^{+3}, S, SO_4^{-2}, and sulfur dioxide, according to the literature (6).

Sodium hydroxide (Eq. 12)

$$8FeS_2 + 30NaOH \rightarrow 4Fe_2O_3 + 14\,Na_2S + Na_2S_2O_3 + 15H_2O \tag{12}$$

has been reported to give a mixture of products in which the sulfur species are sodium sulfide and sodium thiosulfate. Similar reactions have been observed upon treatment of pyrite with Na_2CO_3, $CuCO_3$, Ag_2CO_3, and $Pb(CO_3)_2$.

Stokes (7) observed that mineral pyrite is completely converted into ferrous salt and sulfuric acid by aqueous cupric chloride at $200^\circ C$ (Eq. 13),

$$FeS_2 + 14CuCl_2 + 8H_2O \rightarrow 14CuCl + FeCl_2 + 2H_2SO_4 + 12HCl \tag{13}$$

but copper sulfides are an additional product when cupric sulfate is used. No ferric ion is observed, in accordance with the predictions of the emf series. Silver nitrate is also reported to readily decompose pyrite to ferrous and ferric salts.

Meyers, Hamersma and Kraft (5) recently reported the oxidation of iron pyrite with the very weak oxidizing agent SO_2 in aqueous hydrochloric acid solution. The observed reaction products, Fe^{+2} and elemental sulfur, (Eq. 14)

$$FeS_2 + \frac{1}{2}SO_2 + 2HCl \rightarrow FeCl_2 + \frac{5}{2}S + H_2O \tag{14}$$

are predicted by the standard potentials (Table 3-1). Under the given reaction conditions ($180^\circ C$ solution $0.9\underline{M}$ in H_2SO_4 and $3\underline{M}$ in HCl) 16% reaction was observed in 1 hour, and 55% in 24 hours.

Two reactions involving nonaqueous solvents such as SO_2 and NH_3 should be considered. First, ferric chloride and other metal ions are known to function as oxidizing agents in SO_2 in the conversion of I^- to I_2. Second, oxygen is known to oxidize dissolved metals in ammonia (4). Additional non-aqueous solvents which may be used as media for oxidative (or reductive, cf Section C) reactions are shown in Table 3-2. These solvents have been used as media for electrooxidation and electroreduction with dissolved alkali metals or tetrabutyl ammonium perchlorate electrolyte (8). Presumably electrooxidation or electroreduction of pyrite could be accomplished in solvents such as those listed in Table 3-2. Note that polypropylene carbonate should support the widest range of oxidizing agents without itself undergoing oxidation.

b. Reactions of Ferric Salts

Ferric salts such as chloride, sulfate and ammonium alum are variously claimed by turn-of-the-century literature sources (6) to either *not* attack pyrite or to proceed to 60-80% completion. Interestingly, this controversy has been carried into modern times. Haver and Wong (9) have recently stated (without direct proof) that iron pyrite in copper ore concentrate is not attacked in a period of several hours at $100^{\circ}C$ by aqueous ferric chloride.

Allen (10,6) has claimed that the oxidation of pyrite by relatively con-centrated ferric ammonium alum proceeds to 80.8% conversion at $100^{\circ}C$, and that both sulfate and elemental sulfur are products of the reaction. He reported that the molar ratio of sulfate to sulfur (SO_4^{-2}/S) is 2.4 at low pyrite conversion (and low H^+ concentration) and decreases to 1.5 at high conver-sion (high H^+ concentration). However, if elemental sulfur were formed first (Eq. 15)

$$FeS_2 + Fe_2(SO_4)_3 \rightarrow 3FeSO_4 + 2S \qquad (15)$$

and then subsequently oxidized (Eq. 16),

$$2S + 6Fe_2(SO_4)_3 + 8H_2O \rightarrow 12FeSO_4 + 8H_2SO_4 \qquad (16)$$

the reverse trend to that found by Allen would be obtained. This would predict that the higher the conversion (longer residence time), the more sulfate relative to sulfur (increasing SO_4^{-2}/S ratio) would be produced.

Recent investigations (11) of the leaching of mineral pyrite at $100^{\circ}C$ with relatively concentrated ferric chloride have also shown a trend of decreased

TABLE 3-2

Accessible Potentials of Various Solvents

Solvent	Potential Limits V	
	Anodic	Cathodic
Acetonitrile	2.4	-2.0 - 3.5
Dimethylformamide	1.6 - 2.4	---
Pyridine	1.4	-1.7
Dimethyl sulfoxide	0.7	-1.85
Polypropylene carbonate	5.5 - 5.6	0 - 1.2
Nitromethane	1.7	---
Tetrahydrofuran	1.8 - 4.9	0
Hexamethyl phosphotriamide	1.4	---
Methyl chloride	1.8	-1.7

sulfate-to-sulfur ratio in the products obtained at longer reaction time (Table 3-3) (11), but oxidation proceeds to the 100% level. Elemental sulfur was isolated by extraction of the product with toluene or a chlorinated solvent followed by analysis of the extract for sulfur. Alternatively, the sulfur could be vacuum distilled and collected. Complete dissolution of the pyrite was obtained in 24 hours with the reaction rate decreasing rapidly as the pyrite concentration diminished. Ferric sulfate and ferric nitrate were as effective as chloride (Table 3-3); however, ferric nitrate converted all of the sulfur content to sulfate apparently due to the strong oxidizing action of the nitrate anion (see Table 3-1). Ferric ammonium citrate and oxalate were not very effective, and ferric ammonium sulfate was unstable under the conditions utilized, resulting in precipitate formation without significant reaction.

Garrels and Thompson (12) studied the rate of dilute ferric sulfate (ca. 10^{-4} $\underline{M}$) oxidation at $33^{\circ}C$ of pyrite specimens from three localities. The rate to 50% reduction of the ferric ion by pyrite size fractions between 100 and 200 mesh for the three specimens is shown in Figure 3-1. The stoichiometry was found to be consistent with the following equation:

$$Fe\, S_2 + 8\, H_2O + 14\, Fe^{+3} \rightarrow 15\, Fe^{+2} + 2\, HSO_4^- + 14\, H^+ \qquad (17)$$

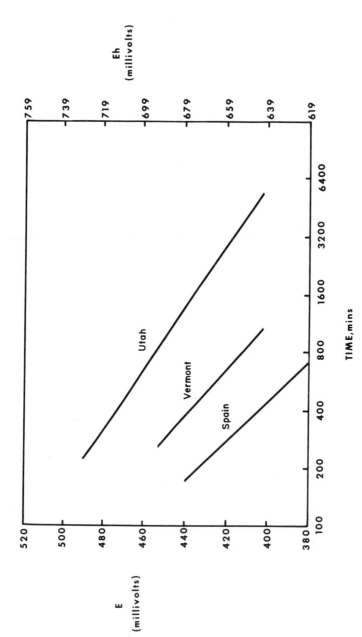

Figure 3-1. Rate of Reduction of Aqueous Ferric Ion by Pyrites
from Three Localities (13)

TABLE 3-3

Extraction of Iron Pyrite at 100°C with Aqueous Ferric Salts*

Ferric Salt	t, hrs	Pyrite Reaction (%)	SO_4^{-2}/S ratio
Cl^-	2	49	3.2
Cl^-	8	96	2.9
Cl^-	16	99	2.5
Cl^-	24	100	2.3
NO_3^-	6	92	All SO_4^{-2}
$SO_4^=$	6	56	—
NH_4 oxalate	6	<10	—
NH_4 citrate	6	<10	—

*-200 mesh mineral pyrite, initial $FeCl_3$ extraction solutions 0.5 $\underline{M}$ in Fe^{+3}, 0.1$\underline{M}^+$; other ferric salts 0.5$\underline{M}$ Fe^{+3}, no added acid.

Elemental sulfur formation (Eq. 15) was not considered in the study of Garrels and Thompson. However, the authors observed that added Cu^{+2} increases the half-life for Fe^{+3} reduction; this was interpreted as a retardation of the rate of pyrite oxidation. However, consideration of the observed oxidation of pyrite by $CuCl_2$ reported by Stokes (7) would lead to the conclusion that $CuSO_4$ was competing in the pyrite oxidation.

Mathews and Robins (13) studied the oxidation of pyrite with somewhat less dilute ferric sulfate (0.12 $\underline{M}$) at 30-70°C (Figure 3-2). They concluded that the rate of ferric ion reduction was first order in ferric ion (Eq. 18)

$$\ln \frac{[Fe^{+3}]}{[Fe^T]} = -Kt \qquad (18)$$

and that the stoichiometry advanced by Garrels and Thompson (Eq. 17) was correct. The effects of pH, total iron, iron species and pyrite surface area were also determined, resulting in an integrated rate expression for reduction of ferric ion (Eq. 19)

$$d\frac{[Fe^{+3}]}{dt} = -K[Fe^{+3}] = -K\frac{WS[Fe^{+3}]}{V[Fe^T][H^+]^{0.44}} \qquad (19)$$

where W = weight of solids in grams, V = volume of liquid involved in reaction in liters and S = specific surface area of FeS_2, M^2/g.

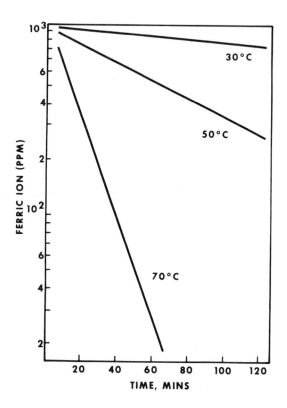

Figure 3-2. Rate of Reduction of Aqueous Ferric Ion
by 36/52 b.s. Mesh Pyrite

The effect of acidity on the rate constant at two particle sizes is shown in
Figure 3-3. The activation energy was found to be 20.5 Kcal/mole. Again, no
attempt was made to isolate any generated elemental sulfur.

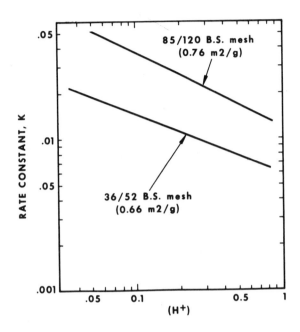

Figure 3-3. Effect of [H$^+$] on Oxidation Rate at 50°C. [Fe^{+3}] = 0.01, Upper Plot and 0.02 Lower Plot

c. Reactions with Oxygen

The air oxidation of pyrite in aqueous media has been studied in order to elucidate the mechanism of acid mine drainage. During coal mining operations, pyrite (in coal) is exposed to air and water releasing dissolved iron and sulfate into the water table.

Acid mine drainage is thought to involve direct oxidation of pyrite or marcasite (Eq. 20)

$$FeS_2 + H_2O + 3\ 1/2\ O_2 \rightarrow Fe^{+2} + 2SO_4^{-2} + 2H^+ \qquad (20)$$

followed by oxidation of Fe^{+2} (Eq. 21) under the catalytic influence of *Thiobacillus ferrooxidans*. As rapidly as produced, Fe^{+3}

$$Fe^{+2} + 1/2\ O_2 + H^+ \rightarrow Fe^{+3} + \frac{1}{2}H_2O \qquad (21)$$

oxidizes additional pyrite (Eq. 22)

$$2Fe^{+3} + FeS_2 \rightarrow 3\ Fe^{+2} + 2S \qquad (22)$$

and the elemental sulfur liberated is oxidized to SO_4^{-2} by Fe^{+3}, or more likely by O_2 catalyzed by *Thiobacillus thiooxidans*, so that elemental sulfur is not reported as a reaction product.

More recent studies (14) on the rate of oxidation of Fe^{+2} and reaction of Fe^{+3} with pyrite under simulated acid mine drainage conditions indicate that Fe^{+3} is the primary pyrite oxidant and that the rate of formation of Fe^{+3} (by oxidation of Fe^{+2}) is the rate determining step.

The possibility of direct reaction of air (oxygen) in aqueous media with pyrite and the role of Fe^{+3} are better understood by consideration of the oxidation of pyrite in dry air. The roasting or air oxidation of pyrite has long been a commercial method for the production of sulfur dioxide to make sulfuric acid. Normally, the product mix consists of various iron oxides and sulfates as well as sulfur dioxide, sulfur trioxide and elemental sulfur; the reaction proceeds at a measurable rate (Figure 3-4) beginning about 300°C (15). By carrying out the reaction isothermally between 400-500°C with abundant air admission and a large specific pyrite surface (16), the pyrite reactions are limited to the following (Eq. 23, 24):

$$FeS_2 + 11/4\ O_2 \rightarrow 1/2\ Fe_2O_3 + 2\ SO_2 \qquad (23)$$

$$FeS_2 + 7/2\ O_2 \rightarrow 1/2\ Fe_2(SO_4)_3 + 1/2\ SO_2 \qquad (24)$$

Under these conditions, nearly 90% of the pyritic sulfur content of mineral pyrite is burned off in 150 minutes at 400°C with only minimal formation of ferric sulfate (Table 3-4). Moreover, the dry environment allows separation

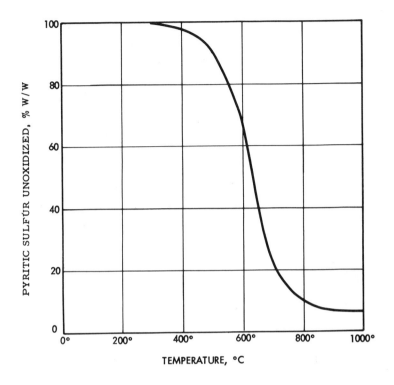

Figure 3-4. Oxidation of Pyrite in Air

TABLE 3-4

Air Oxidation of Powdered* Pyrite at $400^{\circ}C$ (16)

| Time, mins | Rel wt of pyrite dec | | Rel wt unburned pyrite sulfur |
	to Fe_2O_3	to $Fe_2(SO_4)_3$	
10	0.240	0.0402	0.435
50	0.563	0.040	0.267
100	0.846	0.065	0.055
150	0.838	0.097	0.050

*0.01 - 0.1 mm grain size.

of the two variables, Fe^{+3} reaction with pyrite (this must be essentially nil due to the relative immobility of Fe^{+3} in the dry state), and the direct reaction of oxygen. The above results show that air must react very slowly with pyrite below 300-400°C. One must assume that unless water has an unpredictably strong effect on the activation energy of the pyrite oxygen reaction (oxygen is poorly solvated by water), there is no direct reaction of oxygen with pyrite in an aqueous medium at temperatures much below 300°C. Similarly, the rate of oxidation of elemental sulfur by oxygen is not affected by the presence or absence of moisture (15).

McKay and Halpern (17) studied the aqueous oxidation of pyrite with molecular oxygen at temperatures of 100-130°C and P_{O_2} of 1-4 atm where only up to 50% of the pyrite was oxidized. Note, this is considerably less pyrite oxidation than was found under similar conditions in oxidation of pyrite in coal (Chapters 5 and 6). Under all conditions except very low acidity or higher temperatures, a significant buildup of elemental sulfur, formed as an intermediate, was found. The overall kinetic expression is first order in pyrite microscopic surface area and oxygen (Eq. 25) pressure

$$\frac{-d(FeS_2)}{dt} = kA_{FeS_2} P_{O_2} \qquad (25)$$

where $K = 0.125 \exp(-13,000/RT) \text{ m cm}^{-2}\text{atm}^{-1}\text{min}^{-1}$

A_{FeS_2} = microscopic surface area of FeS_2, cm^2

P_{O_2} = partial pressure of oxygen, atm

McKay and Halpern proposed that oxygen, chemisorbed rapidly on the pyrite surface, gives a monolayer of one molecule of O_2 at each FeS_2 site (Eq. 26)

$$FeS_2 + O_2 \text{ (aq)} \xrightarrow{\text{fast}} FeS_2 \cdot O_2 \qquad (26)$$

followed by the slow rate controlling step of O_2 molecules attacking the O_2 covered sites, and decomposition to yield $FeSO_4$ and elemental sulfur (Eq. 27)

$$FeS_2 \cdot O_2 + O_2 \xrightarrow{\text{slow}} (FeS_2 \cdot 2O_2) \xrightarrow{\text{fast}} FeSO_4 + S \qquad (27)$$

and the oxidation of $FeSO_4$ and S to $Fe_2(SO_4)_3$ and H_2SO_4 (Eqs. 28 and 29)

$$4\ FeSO_4 + O_2 + 2H_2SO_4 \rightarrow 2Fe_2(SO_4)_3 + 2H_2O \qquad (28)$$

$$2S + 3O_2 + 2H_2O \rightarrow 2H_2SO_4 \qquad (29)$$

Sato (18) studied the oxidation of a pyrite electrode in aerated iron sulfate solution. He found that below pH = 2, pyrite is in equilibrium with ferrous ions and diatomic elemental sulfur. Abramov (19) reached similar conclusions regarding elemental sulfur formation.

Majima and Peters (20) studied acid leaching of pyrite at $>100^\circ C$, finding that up to 50% of the pyritic sulfur was converted to the elemental sulfur form.

Mathews and Robins (21) studied the reaction of an Australian coal-derived pyrite (about 40% pyrite, 40% marcasite, with quartz making up most of the balance) with molecular oxygen, at $30-70^\circ C$. They found no elemental sulfur at these temperatures. The activation energy, determined at 30 to $70^\circ C$, is 9.3 Kcal/mole (see Figure 3-5) which is somewhat lower than that of McKay and Halpern.

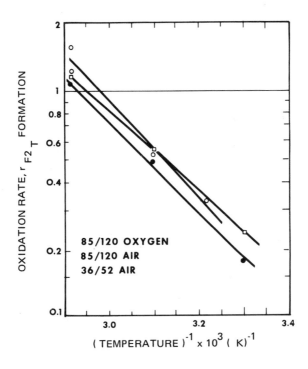

Figure 3-5. Arrhenius Plot, Pyrite Oxidation

In summation:

a)   The oxygen reaction with pyrite is reported to form significant amounts of elemental sulfur at pH of 2 or less (17, 18, 19 and 20) but no elemental sulfur has been identified at low acidity (21).

b)   Ferric ion is generated during the reaction of oxygen with pyrite and is reduced to ferrous ion by reaction with pyrite; but these reactions are considered to be either relatively unimportant (17, 18, 19 and 20), or contrastingly, to be part of the primary mechanism (13).

c)   Ferric ion is variously reported to form about 25% elemental sulfur on reaction with FeS_2 (7,11), or only sulfate (12,13).

d)   A consideration of the direct reaction of O_2 with pyrite in a dry system indicates that the direct reaction of O_2 with FeS_2 should be vanishingly slow at temperatures below 300°C.

Thus, it can be concluded that the mechanism of pyrite oxidation in the presence of oxygen, while not fully resolved, probably involves Fe^{+3}.

C. Reduction of Pyrite

Iron pyrite is reduced by reaction with hydrogen at temperatures somewhat lower than 500°C to form ferrous sulfide and hydrogen sulfide (Eq. 30),

$$FeS_2 + H_2 \rightarrow FeS + H_2S \qquad\qquad (30)$$

but further reduction to elemental iron requires temperatures in excess of 900°C (16). The rate of reduction of FeS_2 by H_2 according to Eq. 30 is shown in Figure 3-6, where a-x = final pyrite weight loss − weight loss at time t. The theoretical pyrite weight loss was 26%, although 24% was obtained at temperatures up to 550°C (16).

Pyrite has the potential to be reduced by aqueous reducing agents (Table 3-5). The metallic reducing agents Mg, Zn, Fe, Sn and Sn^{+2} can be oxidized by FeS_2 provided the kinetics are favorable, although no such reactions are cited in the literature. The possible reduction of pyrite by phosphorus acid (Table 3-5) looks most promising, since both phosphorus acid and phosphoric acid products are water soluble. This approach is limited to the removal of 50% of the pyritic sulfur, since the second step in reduction (to metallic iron) is very difficult in aqueous solution.

Nonaqueous media which may support reduction of pyrite by some of these reducing agents are listed in Table 3-2. Of these, acetonitrile would support the widest range of reducing agent potential.

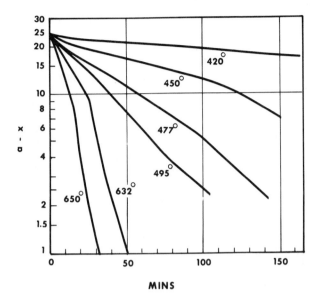

Figure 3-6. Rate of Pyrite Weight Loss at Various
Temperatures in Hydrogen (17)

TABLE 3-5

Standard Reduction Potentials in Aqueous Acid Solution

Half reaction			Potential	
$Mg^{+2} + 2e^-$	=	Mg	-2.37	(22)
$Zn^{+2} + 2e^-$	=	Zn	-0.763	(22)
$2CO_2 + 2H^+ + 2e^-$	=	$H_2C_2O_4$	-0.49	(22)
$Fe^{+2} + 2e^-$	=	Fe	-0.44	(22)
$H_3PO_4 + 2H^+ + 2e^-$	=	$H_3PO_3 + H_2O$	-0.276	(22)
$N_2 + 5H^+ + 4e^-$	=	$H_2NNH_3^+$	-0.23	(22)
$Sn^{+2} + 2e^-$	=	Sn	-0.136	(22)
$H^+ + e^-$	=	$1/2\ H_2$	+0.00	(22)
$Sn^{+4} + 2e^-$	=	Sn^{+2}	+0.070	(22)
$FeS_2 + 2H^+ + 2e^-$	=	$FeS + H_2S$	+0.165	(2)

II. DESULFURIZATION OF ORGANIC SULFUR COMPOUNDS

Organic sulfur is thought to occur in four major forms (Chapter 2). These are mercaptans or thiols, sulfide, disulfide, and aromatic ring sulfur as exemplified by the thiophene system. The chemistry of organically bound sulfur compounds of the type thought to be present in coal has been reviewed by Given and Wyss (23), and excellent comprehensive treatises on the chemistry of organic sulfur are available in the literature (24, 25, 26). The following sections review the literature from the standpoint of reactions which can be expected to result in either 1) desulfurization of organic sulfur compounds of the type found in coal, or 2) partition and removal of organic sulfur compounds from the coal macromolecule.

A. Classification of Desulfurization Methods

Potential desulfurization methods for coal may be classified into six groups: solvent partition, thermal decomposition, acid base reaction, reduction, oxidation and displacement.

1. SOLVENT PARTITION

The organic portion of the coal system is soluble in varying degrees in a vast number of organic solvents which can dissolve essentially all of the organic matrix, depending on the specific solvent used and the temperature of extraction (27). In order for "desulfurization" to take place as a result of the solvent extraction of coal, the solute must contain a higher ratio of organic sulfur to carbon than the residue. The higher this ratio, the higher the desulfurization selectivity.

In order to enhance the solubility of the sulfur-containing coal molecules, it could be possible to add coordinating agents to the solvent which would form soluble salt-like complexes (Eq. 31) which could be

$$R_1 S_X R_2 + X \rightarrow \begin{matrix} R_1 \\ \\ R_2 \end{matrix} \!\!\!> \overset{+}{S}_X \quad X^- \qquad (31)$$

decomposed after extraction for recovery and recycle. The selectivity of solvent partition of organic sulfur would be enhanced if the coal macromolecule could be cleaved such that the organic residues associated with dissolved organic sulfur (i.e., R_1 and R_2) are relatively low in molecular weight.

2. THERMAL DECOMPOSITION

As previously noted, coal begins to decompose rapidly at temperatures near 400°C. Therefore, within the context of this discussion (which has as its objective desulfurization while maintaining the coal yield in a condition as near to its virgin state as possible), thermal decomposition of organic sulfur compounds (Eqs. 32, 33) would have to take place below 400°C and the

$$R_1 S_X R_2 \overset{\rightarrow}{\Delta} R_1 R_2 + S_X \tag{32}$$

$$RCH_2 CH_2 SH \overset{\rightarrow}{\Delta} RCH = CH_2 + H_2S \tag{33}$$

cleavage of sulfur would have to occur in preference to or predominate over the cleavage of other coal compounds. While organic sulfur compounds are subject to pyrolysis, so are the other components of the coal macromolecule such as amines, ethers, acids, etc., and C-O bonds are more easily broken than C-S bonds (23). Therefore, a catalytic approach to pyrolytic decomposition, in which a catalyst selectively promotes the decomposition of organic sulfur, is the only possibility likely to be successful.

3. ACID-BASE NEUTRALIZATION

Mercaptans or thiols, which are weakly acidic, are capable of being neutralized (Eq. 34) and brought into aqueous solution by strong bases.

$$R S H + O H^- \rightarrow R S^- + H_2O \tag{34}$$

This is not true of sulfides, disulfides or aromatic sulfur compounds. Thus, a caustic leaching process limited to acid-base neutralization would remove only thiols. As in solvent partition, a preliminary lowering of the molecular weight of the coal macromolecule would aid the selectivity of this type of treatment.

4. REDUCTION

Organic thiols, sulfides and disulfides, as well as aromatic organic sulfur, are susceptible to reduction by molecular or nascent hydrogen (Eq. 35) or hydrogen donor systems (Eq. 36).

$$R_1 S_X R_2 + 4H \rightarrow R_1 H + R_2 H + H_2 S_X \tag{35}$$

$$R_1 S_X R_2 + 2 R_3 H \rightarrow H_2 S_X \tag{36}$$

Note that sulfides and thiols would give hydrogen sulfide as a product, while under mild reduction conditions disulfides would give hydrogen polysulfides (under more vigorous-reducing conditions, the product would be H_2S). Hydrogen polysulfides are thermally unstable and would tend to dissociate into hydrogen sulfide and elemental sulfur. Any elemental sulfur which survived the reducing treatment could then be removed from the reactants by careful vaporization or by extraction with a solvent selective to sulfur. The desulfurization of coal with hydrogen at temperatures above $400^{\circ}C$ is well known (coal liquefaction) and gives rise to liquid and gaseous hydrocarbon products. Again, within the present context, discussion is limited to reduction at lower temperatures where selectivity may be enhanced and coal would maintain its identity as a solid.

5. OXIDATION

Organic thiols, sulfides and disulfides may be oxidized to sulfones, then to sulfoxides and finally to sulfonic acids (thiols are first oxidized to disulfides). The sulfonic acids can then be hydrolyzed in boiling water to eliminate sulfuric acid as the sulfur product (Eq. 37) which can be

$$R_1 S_X R_2 \underset{[O]}{\rightarrow} R_1 SO_3 H + R_2 SO_3 H \underset{\Delta}{\overset{H_2O}{\rightleftharpoons}} R_1 O H + R_2 O H + 2 H_2 SO_4 \quad (37)$$

filtered away from the solid coal matrix. Thus, at every former sulfur position which has been oxidized and hydrolyzed, an alcohol functional group would remain. The key to a desulfurization process based on oxidation would be the minimization of the oxidation of the remainder of the coal matrix, which is a difficult task since amines, alcohols, esters and olefinic groups are as easily oxidized as sulfur-containing units.

6. NUCLEOPHILIC DISPLACEMENT

Although many sulfur compounds are nucleophilic, a strong nucleophile can attack the sulfur atom (Eq. 38) of a sulfide or disulfide giving an alkyl (Eq. 38) or thioalkyl (Eq. 39) leaving group or the nucleophile can attack a carbon atom bonded to a sulfide group (Eq. 40). In each case sulfur is

$$R_1 S_X R_2 + Nu^- \rightarrow R_1 S_X Nu + R_2^- \quad (38)$$

$$R_1 S_X R_2 + Nu^- \rightarrow R_1 S_{X-1} Nu + R_2 S^- \quad (39)$$

$$R_1 S_X R_2 + Nu^- \rightarrow R_1 S_X^- + R_2 Nu \quad (40)$$

cleaved from the coal matrix in reactions 38 and 40 if R_2 is the bonding point with the matrix residue and in reaction 39 if R_2 is a side-chain. These reactions can be visualized as sulfur removal methods if the leaving group contains few carbon atoms relative to the number of cleaved sulfur atoms and is either soluble or volatile, so that the sulfur-containing moiety can be filtered or volatilized away from the coal matrix. A secondary reaction can also be envisioned in which the displaced fragment (from Eq. 39) now acts as a nucleophilic agent attacking the carbon atom adjacent to the sulfur function (Eq. 41) so that the nucleophile now has become a sulfur

$$R_2 S^- + R_1 S_{X-1} Nu \rightarrow R_1 S_X R_2 + Nu S \tag{41}$$

scavenger. In this case, sulfur would be removed from the system attached to the nucleophile provided the combined molecule is either soluble or volatile.

B. Desulfurization Reactions

Specific examples of the above-mentioned potential coal desulfurization reactions, taken from the literature, are presented in the following sections. The majority of the references stem from petroleum refining literature which is particularly applicable since organic sulfur is present in petroleum in analogous structural forms (Chapter 2) and since chemical desulfurization has always been an important oil refining method.

1. SOLVENT PARTITION

Powell (28) investigated the phenol dissolution of organic sulfur fractions from several coals. From 4 to 43% of the organic sulfur content of these coals was found to dissolve in phenol at 150°C over a 20-hour period. The results for the six coals investigated are plotted in Figure 3-7 as a function of fixed carbon content. As indicated in the figure, the lower the fixed carbon, the larger the percentage of organic sulfur which is dissolved.

Unfortunately, since Powell did not obtain a material balance for this experiment, the selectivity of his phenol dissolution is not known. Thus, it is possible that the ratio of organic sulfur to carbon in the coal extract is the same as the ratio in the residue in each case. It is therefore not known whether desulfurization had indeed taken place. Very likely, unless phenol has a strong depolymerizing action on the coal, the carbon groups attached to the sulfur atoms in the solute would likely be large enough to eliminate any strong selectivity. Wheeler (29) reported that pyridine and chloroform coal extracts had no increase in organic sulfur.

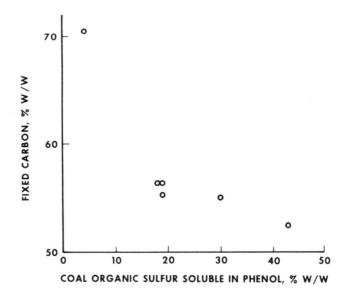

Figure 3-7. Extraction of Organic Sulfur from Illinois Coals

Fisenko and co-workers (30) reported that both pyridine and phenols cause some destruction of the coal macromolecule, with part of the organic sulfide compounds being transferred into solution. However, thiol groups were found to be mostly insoluble, and therefore bound firmly to the main coal macromolecule. It has recently been shown (31) that phenol containing p-toluene sulfonic acid catalytically decomposes the organic coal matrix of some Japanese coals to give molecules of 300-1100 molecular weight soluble in the typical coal extraction solvents pyridine and benzene/ethanol. However, no information was given regarding the fate of organic sulfur compounds in these depolymerization-dissolution systems. It was also found that phenol itself had only a small depolymerizing action.

Heredy and Neuworth (32) found that boron trifluoride-phenol at 100°C extensively depolymerizes coal to an average molecular weight of 425 but leaves a large residual phenol, or phenol-derived material, residue with the coal. The methanol-soluble fractions were found to have 50% higher sulfur content than the other fractions but constituted only 1.8% of the extracted coal.

It has been reported that diethyl sulfide dissolves in liquid hydrofluoric acid giving the corresponding salt which can be decomposed by addition of water to recover the sulfide (33, 34). Similarly, lower alkyl sulfides can be extracted from hydrocarbon solutions by liquid hydrofluoric acid (34) or by concentrated sulfuric acid (33). Anhydrous HF is reportedly effective (35) in extracting organic sulfides, disulfides and thiophenes from fuel oil at moderate temperatures (100-350°F). Significantly, the HF extract comprises 20-40% of the fuel oil treated so that selectivity is not as high as desired. Sulfuric acid has also been used (36,37) for the removal of organic sulfur compounds from medium distillates and for the refining of crude oil. Liquid sulfur dioxide at temperatures below its boiling point (<8°C), has been used for refining crude oil in a number of installations (38). Here aromatic and unsaturated hydrocarbons, as well as a large proportion of the sulfur compounds present in the feed, are dissolved with little evidence of chemical reaction. The sulfur dioxide is recovered easily by distillation leaving bottoms high in sulfur content.

Thus, a sufficiently depolymerized coal could conceivably be extracted with liquid hydrofluoric acid, sulfur dioxide or aqueous sulfuric acid to selectively dissolve sulfides, disulfides and possibly some aromatic sulfur compounds, which could be recovered on dilution with water, adsorption on silica or alumina gel, distillation of the solvent or freezing out. A negative aspect of the use of sulfuric acid lies in potential sulfonation and oxidation of the coal matrix.

Coal depolymerization could be effected by p-toluene sulfonic acid in phenol as noted above, by mild hydrogenation (at temperatures below 400°C and in the absence of solvent), a prepyrolysis, oxidation, or simply by increasing the extraction temperature.

The subject of pretreatment of coal for enhancement of solubility is comprehensively treated by Dryden (39). Preheating in an inert atmosphere for several hours at temperatures of 180 to 400°C has been shown to increase the solubility of several coals toward solvents such as pyridine and benzene, with up to a seven-fold increase of solubility in chloroform noted for extraction of a coal preheated at 330 to 340°C for 22 hours. Oxidation is variously reported to increase, decrease or not affect the solubility of coal while partial hydrogenation strongly increases the benzene and pyridine extraction yields for the coals investigated.

2. THERMAL DECOMPOSITION

Most organic sulfur compounds do not undergo thermal decomposition below the 400°C thermal decomposition temperature of coal. Ethyl sulfide begins decomposition at 400°C and many mercaptans decompose to olefins and H_2S at around 300°C; however, most organic sulfur compounds require temperatures in the neighborhood of about 475°C (24) even under the influence of catalysts. Dibenzyl sulfide is a special case, decomposing at 260°C and forming stilbene, hydrogen sulfide, and other products. However, diphenyl sulfide, under similar conditions, gives dibenzothiophene, hydrogen sulfide and benzene. Thus, the former compound can be pyrolytically desulfurized, while the latter would result in a partial removal of sulfur as hydrogen sulfide on pyrolysis. A great deal of experimentation has been performed on the pyrolysis of coal, and sulfur compounds volatile at 100 to 300°C have been detected (40), but no selective decomposition of the organic sulfur portion of the coal macromolecule has been reported. Hence, unless a degree of selectivity can be introduced through use of a solid or liquid catalyst, there appears to be no particular promise in this area of desulfurization. Liquid catalysts which may prove effective include phosphoric acid and cresol which catalyze decomposition of mercaptans (24).

3. ACID-BASE NEUTRALIZATION

The use of aqueous alkali to remove organic mercaptans from petroleum fractions is well-known technology (37). As the molecular weight of the mercaptan increases, its solubility in alkali decreases until only 2% of C_7 mercaptans are removed with 10% aqueous alkali and 73% with 40% caustic. Increased solubility can be obtained by the addition of organic solvents such as methyl alcohol or organic acids. Mercaptans thus extracted from petroleum fractions are alkali metal salts, which may be directly oxidized to disulfides with air, regenerating caustic and forming an immiscible alkyl disulfide phase which is easily separated. Thus, to the extent that free thiols are present in coal and the molecular weight does not exceed the limits of the caustic extraction media, coal may be desulfurized at moderate temperatures (usually 100°F) with aqueous sodium hydroxide, for example, and then can be recycled and separated from the solution with regeneration, as noted above.

It would seem possible to increase the thiol content of coal by mild reduction with hydrogen to cleave some sulfides and disulfides to thiols. Here again, depolymerization of the coal matrix would materially aid in the selectivity of any acid base neutralization extractions. Further, mild pyrolysis could decompose some sulfides and disulfides to thiols while depolymerizing the coal matrix.

4. REDUCTION

Catalytically activated reduction of non-aromatic sulfides and disulfides begins at about 250°C to form the cleavage products thiol and alkane (24). Some exceptional catalysts, such as Raney nickel or cobalt-molybdenum, can remove sulfur even from aromatic sulfur ring compounds to give hydrogen sulfide. However, extensive studies on the hydrogenation of coal have shown that temperatures of 400°C or more and the presence of a hydrogen donor solvent are needed for removal of more than a small fraction of the organic sulfur, even in the presence of a catalyst.

Alkali metals, such as sodium and potassium, are capable of cleaving sulfides and disulfides (24) but only to the thiol salt stage, hence no removal of sulfur can be obtained. Phosphorous and phosphine should be capable of reducing organic sulfur to form phosphorous polysulfide and hydrogen sulfide, respectively. Since both phosphorous and most phosphorous polysulfides are soluble in organic solvents, it would be possible to contact the reagent with coal and remove the product sulfur compound. The polysulfides of phosphorous react with water to form phosphorous oxyacids and H_2S. The oxyacids can be reduced with carbon (coal) to regenerate phosphorous. Phosphorous polysulfide is used in a mixture with sulfur for sulfuration of organic oxygen compounds (26). The use of a large excess of phosphorous could minimize this "back reaction," however, and allow removal of organic sulfur.

5. OXIDATION

The action of sodium hypochlorite on thiols, disulfides and sulfides (41) show that these compounds are oxidized in a matter of minutes at room temperature to sodium salts of sulfonic acids. Salts of the lower alkyl groups were determined to be soluble. However, in the case of coal, the use of sodium hypochlorite could lead to problems, since this media would tend to chlorinate and oxidize the coal matrix.

Dilute warm nitric acid treatment of coal has been found to remove some organic sulfur by oxidation (42), while concentrated nitric acid attacks the organic sulfur compounds of coal even at room temperature. However, oxidation and nitration by nitric acid is always excessive under conditions in which organic sulfur is oxidized (43), so that little or no selectivity is likely. Other oxidizing agents capable of oxidizing organic sulfur compounds to soluble sulfonic acids include hydrogen peroxide, ozone, various oxides of nitrogen, and permanganate.

If petroleum asphaltenes may be considered as a model for the organic portion of the coal matrix, the results of oxidation of three petroleum asphaltenes may be of interest.

The elementary composition of three petroleum asphaltenes oxidized with alkaline permanganate shows selectivity for sulfur removal in one case (Table 3-6). As indicated by the data, only a small part of the asphaltene was dissolved in sample 3, so that the selectivity for sulfur oxidation was fairly high. However, in all cases the residue was much higher in oxygen than the native material, which is indicative of unselective oxidation.

TABLE 3-6

Oxidation of Asphaltenes with Permanganate* (44)

Sample	Asphaltene residue (%)	Elementary composition (%)				
		C	H	O	N	S
1. Lagunillas						
Native	––	84.2	7.9	1.6	2.0	4.5
Residue	37	76.5	7.1	10.4	0.7	5.2
2. Burgan (Kuwait)						
Native	––	82.2	8.0	0.6	1.7	7.6
Residue	89	76.6	7.3	7.9	0.8	7.3
3. Baxterville						
Native	––	84.5	7.4	1.7	0.8	5.6
Residue	86	83.5	7.5	4.2	0.8	3.9

*100 hrs at room temperature

A comparison of half-wave potentials for oxidation of organic substrates of the type found in coal (Table 3-7) shows that: 1) alcohols and amines are vastly more susceptible to oxidation than ethers and benzene derivatives, and 2) the range of oxidation potentials is large. Presumably, the oxidation potential of organic sulfur compounds would lie roughly in the same region as amines and alcohols. Thus, given a mild oxidizing agent, it might be possible to oxidize the organic sulfur compounds, amines and alcohols without destroying the hydrocarbon portion of the coal molecule.

TABLE 3-7

Oxidation Potentials of Organic Substrates of the Type Found in Coal (45)

Compound	E vs. SCE (V)
Mesitylene	1.90
Anisole	1.67
Naphthalene	1.72
Anthracene	1.20
Aniline	0.72
Phenol	0.633
p-Methoxyaniline	0.44

6. NUCLEOPHILIC DISPLACEMENT

An interesting double nucleophilic substitution (31) involves the displacement of sulfur from disulfides by phosphine or phosphinate (Eq. 42),

$$R\text{-}S\text{-}S\text{-}R + R_3\,P \;\rightarrow\; \left[R_3\,\overset{+}{P}\!-\!S\!-\!R \atop RS^- \right] \;\rightarrow\; R_3\,\overset{\displaystyle S}{\overset{\displaystyle \|}{P}} + R\text{-}S\text{-}R \qquad (42)$$

which is analogous to a combination of Eqs. 39 and 41 described in the previous section. Here, half of the sulfur could be removed from disulfides by phosphine or phosphinate, as discussed above.

The attack of caustic on sulfides or disulfides to yield alcohols and hydrogen sulfides (Eq. 43) would be a nucleophilic substitution. These reactions

$$R\text{-}S_X\text{-}R + Na\,O\,H \;\rightarrow\; R\,S_X\,Na + R\,O\,H \overset{Na\,O\,H}{\rightarrow} Na_2\,S_X + 2\,R\,O\,H \qquad (43)$$

are not reported in organic chemistry literature, although they seem feasible at high temperatures. Similarly, the attack of hydrogen iodide (or hydrogen bromide) on a thiol to form an alkyl iodide and hydrogen sulfide (Eq. 44)

$$R\,S\,H + H\,I \;\rightarrow\; R\,I + H_2S \qquad (44)$$

would result in desulfurization. The alkyl iodide could then be hydrolyzed in water to the alcohol and hydrogen iodide recovered from the aqueous filtrate.

REFERENCES

1. R.J. Biernat and R.G. Robins, Electrochim Acta, 14: 809 (1969).

2. R.J. Biernat and R.G. Robins, Electrochim Acta, 17: 1261 (1972).

3. W.M. Latimer, The Oxidation States of the Elements and Their Potentials in Aqueous Solutions, Prentice-Hall, New York, 1938.

4. E.S. Gould, Inorganic Reaction, Holt, Rinehart and Winston, Inc., New York, 1962.

5. R.A. Meyers, J.W. Hamersma and M.L. Kraft, Environ. Sci. and Techn., 9: 70 (1975).

6. J.W. Mellors, (ed.), A Comprehensive Treatise on Inorganic and Theoretical Chemistry, Vol. 14, Wiley, New York, 1961.

7. H.N. Stokes, Econ. Geol, 2: 14 (1907).

8. S.D. Ross, M. Finkelstein and E.J. Rudd, Anodic Oxidation, Academic Press Inc., New York, 1975, pp 74-76.

9. F.P. Haver and M.M. Wong, J. Met., 23: 25 (1971).

10. E.T. Allen and J.L. Crenshaw, Amer. J. Sci, 38: 393 (1914).

11. J.W. Hamersma, E.P. Koutsoukos, M.L. Kraft, R.A. Meyers, G.J. Ogle, and L.J. Van Nice, Chemical Desulfurization of Coal: Report on Bench-Scale Developments, Vol. 1, 1, EPA-R2-73-173a, U.S. Government Printing Office, Washington, D.C., 1973.

12. R.M. Garrels and M.E. Thompson, Am. J. Sci., Bradley Volume, 258-A: 57 (1960).

13. C.T. Mathews and R.G. Robins, Austral. Chem. Eng., 21, August 1972.

14. P.C. Singer and W. Stumm, Oxidation of Ferrous Iron, Water Pollution Control Research Series, DAST-28, 14010-06/69, U.S. Dept. of the Interior, Washington, D.C., 1969.

15. J.W. Mellors, (ed.) A Comprehensive Treatise on Inorganic and Theoretical Chemistry, Vol. 10, Wiley, New York, 1961.

16. G.M. Schwab and J. Philinis, J. Am. Chem. Soc., 69: 2588, (1947).

17. D.R. McKay and J. Halpern, J. Trans. Metallurgy Soc. of A.I.M.E., 212: 301 (1958).

18. M. Sato, Econ. Geol., 55: 1202 (1960).

19. A.A. Abramov, Ysvetnye Metally., 33 (Dec. 1965).

20. H. Majima and E. Peters, Proc. I.U.P.A.C. Conference, Sydney, Australia (1969).

21. C.T. Mathews and R.G. Robins, Austral. Chem. Eng., 19 (Nov/Dec 1974).

22. J.A. Dean, (ed.) Lange's Handbook of Chemistry, McGraw-Hill Book Company, New York.

23. P.H. Given and W.F. Wyss, British Coal Utilization Research Association Bull., 25: 165 (1961).

24. E.E. Reid, Organic Chemistry of Bivalent Sulfur, Chemical Publishing Co. Inc., New York, 1960.

25. A. Senning, ed., Sulfur in Organic and Inorganic Chemistry, Vol. I, Dekker Inc., New York, 1971.

26. N. Kharasch, Organic Sulfur Compounds, Pergamon Press, New York, 1961.

27. W.S. Wise, Solvent Treatment of Coal, Mills and Boon Ltd., London, 1971.

28. A.R. Powell, J. Ind. and Eng. Chem., 12: 887 (1920).

29. R.V. Wheeler, Coll. Guard, 121: 1596 (1921).

30. N.N. Fisenko, V.A. Larina and A.D. Baranskii, Vop. Khim. Khim. Tekhnol, 141 (1969) Chem. Abstr., 72, 23250g (1970).

31. K. Ouchi, K. Imuta and Y. Yamashita, Fuel (London) 44: 205 (1965).

32. L.A. Heredy and M.B. Neuworth, Fuel, 41: 221 (1962).

33. J.R. Meadow and T.A. White, Ind. Eng. Chem., 42: 925 (1950).

34. A.P. Lien, D.A. McCaulay and B.L. Evering, Ind. Eng. Chem., 41: 2698 (1949).

35. C.N. Kimberlin, Jr., U.S. Patent 3,383,300 (1968).

36. Y.B. Chertkov, Khim. Seraorg. Soedin. Soderzh. Neftyakh Nefteprod., 8: 387, 1968. Chem. Abstr., 71, 83205k.

37. W.A. Gruse and D.R. Stevens, Chemical Technology of Petroleum, McGraw-Hill Book Co. Inc., New York, 1960, pp 302-304.

38. W.A. Gruse, Petroleum and its Products, McGraw-Hill Book Co. Inc., New York, 1928.

39. I.G.C. Dryden, Brit. Coal Util. Res. Assoc. Bull, 13: 113 (1949).

40. A. Lissner and A. Nemes, Brennst. Chem., 16: 101 (1935).

41. S.F. Birch, W.S. Gowan and P. Norris, J. Chem. Soc., 127: 1934 (1926).

42. R.A. Mott, Fuel, 29: 53 (1959).

43. B.K. Mazumdar, A.K. Chatterjee and A. Lahiri, Fuel (London), 46: 379 (1967).

44. J.G. Erdman, in Geochemistry of the High Molecular Weight Non-Hydrocarbon Fractions of Petroleum, Advances in Organic Geochemistry, (V. Columbo and G.D. Hobson, eds.), MacMillan, New York, 1964.

45. S.D. Ross, M. Finkelstein and E.J. Rudd, Anodic Oxidation, Academic Press Inc., New York, 1975, pp 84, 190 and 272.

CHAPTER 4

CRITERIA FOR SUCCESSFUL CHEMICAL
DESULFURIZATION PROCESSES

A number of chemical reagents were shown in Chapter 3 to be candidates for the chemical removal of pyritic and organic sulfur from coal. These desulfurization chemicals include: a) metals — zinc, iron, tin, and metalloids such as phosphorus, b) metal salts — ferric sulfate, ferric chloride, cupric sulfate, c) acids — hydrogen iodide, hydrogen bromide, hydrogen fluoride, p-toluene sulfonic acid, and oxalic acid, d) bases — sodium hydroxide and lime, e) oxidizing and reducing agents — hydrazine, hydrogen peroxide, etc., and f) solvents — cresol, phenol, sulfur dioxide and ammonia. In each case, one or more chemicals are consumed and one or more chemicals are formed by the desulfurization reaction. The number of desulfurization chemicals is large so that some criteria must be established for the sorting and evaluation of promising chemical desulfurization approaches for further investigation.

The selected criteria may include the following important considerations: a) the reagent must be highly selective to either the pyritic or organic sulfur content of coal (or both) and not significantly reactive with other coal components, b) the reagent must be regenerable so that once-through reagent cost is not a major factor, c) the reagent should be either soluble or volatile in both its unreacted and reacted form so that it can be near totally recovered from the coal matrix, and d) the reagent should be inexpensive, since a portion of it will certainly be lost either to irreversible sorption on the coal matrix or consumed in some other way.

The market cost of a chemical reactant for chemical desulfurization is an important factor even if the reagent is to be regenerated. Often the market cost is a good approximation of the regeneration expense. The cost of a number of reagents which have been suggested in Chapter 3 is shown in Table 4-1. Not included in the table are oxygen and hydrogen, which are key potential desulfurization chemicals, since they are commodity chemicals at the low end of the tabulated cost range. Consider first the situation where about 1 percent by weight reagent is irreversibly sorbed into the coal matrix (see Chapter 2, where the porosity of coal was discussed) thus destroying 20 pounds of reagent per ton of coal. In the case of hydrogen iodide, this would amount to a $90/ton of coal cost to the process, which would increase the cost of product steam coal by about 4 to 5 times over its present value. Obviously, the same situation for cresol would be much less deleterious, increasing the cost of product coal by about $9/ton, though still a problem. For sodium hydroxide or ferric sulfate, the increase in cost would be only $1.40 and $0.60/ton of coal, respectively.

From a pollution control standpoint, the irreversible sorption of some of the reagents e.g., hydrogen iodide, hydrogen bromide, hydrogen fluoride which have been suggested would be unacceptable. These chemicals would give rise to a new pollution problem (halogen) on combustion in a utility boiler, as well as promote boiler corrosion. Residual sodium hydroxide would not be a pollution problem but would corrode utility boilers. However, the retention of 1 percent sulfur dioxide on coal would only lower the desulfurization efficiency, leaving about 0.5 percent w/w sulfur in the coal. It is tolerable, from a pollution control standpoint, to leave residual chemicals on coal which contain only those elements normally present in the coal. Reagents of this type include cresol, phenol, oxalic acid (carbon and oxygen), ammonia and hydrazine (nitrogen and hydrogen), hydrogen peroxide (hydrogen and oxygen), etc., as well as lime (calcium and oxygen).

TABLE 4-1

Cost of Desulfurization Chemicals (1)

Desulfurization chemical	Cost of chemical/lb ($)
Calcium oxide (lime)	0.01
Ferric sulfate	0.03
Sodium hydroxide (caustic soda)	0.07
Sulfur dioxide	0.07
Ammonia	0.09
p-Toluene sulfonic acid	0.25
Hydrogen peroxide	0.26
Phenol	0.26
Phosphorus	0.28
Oxalic acid	0.32
Hydrogen fluoride	0.41
Cresol (cresylic acid)	0.44
Hydrogen bromide	0.75
Hydrazine	1.60
Hydrogen iodide	4.50

The desulfurization reagent should be either soluble or volatile in both its native and reacted form. Some examples of reagents which are soluble and give soluble and/or volatile products are shown in Table 4-2. Alternatively, it may be possible to recover the reaction products by subsequent reaction. For example, in the cleavage of organic sulfides or disulfides by halogen acids, the product aryl and alkyl halides could be decomposed by hydrolysis to regenerate the halogen acids.

TABLE 4-2

Desulfurization Reagent — Product Pairs

Reagent	Products
H_3PO_3	H_3PO_4, H_2S
$Fe_2(SO_4)_3$	$FeSO_4, H_2SO_4, S_x$
H_2NNH_2	N_2, H_2S
$H_2C_2O_4$	CO_2, H_2S
P	P_xS_y
O_2	$FeSO_4, H_2SO_4, S_x$
Cresol, P-toluene sulfonic acid	Same (catalyst only)

Finally, it is highly important that the desulfurization reactions be quite selective to the sulfur content of the coal. Destruction of the coal by oxidation or reduction or total dissolution in a partitioning solvent can destroy the cost effectiveness of the entire process.

REFERENCES

1.   Chemical Marketing Reporter, April 5, 1976.

CHAPTER 5

PYRITIC SULFUR REMOVAL PROCESSES —
METAL ION OXIDANTS

I. Introduction

II. Chemistry and Process Data
   A. Selection of Specific Reagents and Conditions
      1. Process Description
      2. Selection of Specific Reagents and Conditions
   B. Rate of Pyritic Sulfur Removal
   C. Regeneration of Ferric Sulfate Leach Solution
   D. Simultaneous Pyrite Leaching and Ferric Sulfate Regeneration
      1. Selection of Conditions
      2. Initial Engineering Data
      3. Advanced Engineering Data

III. Recovery of Desulfurization Products
   A. Removal of Elemental Sulfur from Coal
      1. Vaporization
      2. Solvent Extraction
      3. Chemical Reaction
   B. Rejection of Product Iron and Sulfate
   C. Fate of Minor and Trace Elements

IV. Applicability to U.S. Coals
   A. Sulfur Removal
   B. Selectivity and Heat Content Changes

V. Engineering Design and Cost Estimations
   A. TRW Engineering Studies
      1. Suspendable Coal Processing Design and Cost Studies
      2. Coarse Coal Processing Design and Cost Studies
      3. Projection of Process Economics
   B. Dow Chemical — USA Design and Cost Estimation Studies
      1. Process Design
      2. Process Economics
      3. Process Byproducts
   C. Exxon Research and Engineering Co. Design and Pollution
      Control Studies

I. INTRODUCTION

A number of chemicals which have desulfurization reactivity with functional groups of the type found in coal were discussed in Chapter 3. Several key engineering and economic criteria for the development of a viable coal desulfurization process were described in Chapter 4. The remaining six chapters will review the methods which have been advanced for the chemical desulfurization of coal and will propose some new methods, with special emphasis on engineering and economic viability of the total process, particularly in rela-

59

tion to sulfur removal, regeneration of the reagent and disposal of the sulfur derived reaction products.

There is a nearly total absence of reported research on processes for the chemical desulfurization of coal prior to the flurry of very recent activity stimulated by the advent of air pollution control legislation and resultant Federal interest in methodology for control of sulfur oxide pollution. In fact, up until the advent of the ferric ion-based leaching process, which is discussed in the following sections, chemical desulfurization was considered to be at best a formidable problem and at worst an impossible task. However, now that this in a sense pioneering process has been thoroughly investigated and found to be promising, the concept of chemical desulfurization no longer appears to be a "blue sky" proposition and some additional processes have lately received attention. Furthermore, as I will attempt to demonstrate in the following discussion, the field may indeed be wide open to the intrepid and innovative investigator.

As indicated in Chapter 3, a number of metal ions have sufficient oxidation potential to oxidize pyrite, e.g., Hg_2^{+2} and Ag^+. However Hg_2^{+2} and Ag^+ deposit the free metal as a reaction product which would not only contaminate the coal, but would have a nonrecoverable cost determined by the weight of reacted metal. The utilization of $CuCl_2$ would seem feasible since it has been demonstrated effective for leaching mineral pyrite. Possible disadvantages lie in the contamination of coal with Cl^- and separation of generated iron salts from the copper-based leach liquor. Apparently, the utilization of $CuSO_4$ would result in deposition of some copper sulfides which would lower the efficiency of sulfur removal and contaminate the coal with copper.

Thus, in general, the use of metal ions for removal of pyritic sulfur from coal is open to question. However, a process based on the oxidation of coal pyrite with a metal ion is being seriously advanced as a means of producing desulfurized coal. This is the ferric ion leaching system, known as the Meyers Process (1). Once having established a reasonable set of criteria for selection of an economically viable chemical desulfurization process (Chapter 4), and having knowledge of the reactivity of ferric ion with mineral pyrite (Chapter 3), (although the literature was conflicting), it would appear obvious that ferric ion leaching of coal should provide a promising basis for chemical desulfurization. More specifically: a) ferric salts, although effective pyrite oxidants, should be reasonably selective to pyrite as the oxidation potential of the ferric couple is relatively weak (cf. Table 3-1), b) ferric ion is readily regenerable with air or oxygen, c) several of the ferric-ferrous salt pairs are highly soluble in water, d) the products of the pyrite-ferric ion reaction, sulfur and iron sulfate, are relatively environmentally innocuous and storable, and e) in the use of ferric sulfate, there is no chance of contamination of the coal, since iron sulfates are a natural coal component.

Astoundingly, the process was undiscovered until 1970 (2). On consideration, three factors may have played a role here: 1) conflicting literature at that
time (Chapter 3-I.B.2.b.) as to whether ferric ion does indeed oxidize pyrite and
to what extent, 2) a relative paucity of interest in pollution control and hence
coal desulfurization and 3) availability of literature (3) which indicated that
elemental sulfur (a product sometimes associated with the reaction of pyrite
with ferric ion) reacts with coal, which led to the inference that sulfur would
not be removable and any process based on ferric ion leaching would be relatively ineffective.

Following convincing exploration of the process chemistry (4-8) for
removal of pyritic sulfur from coal utilizing ferric salts, a concerted effort by
several engineering organizations was initiated to gather the extensive data and
engineering design experience necessary to make the Meyers Process available
for construction of full-scale desulfurization plants. A critical summary of the
available information is presented in the following four sections, which are
entitled: Chemistry and Process Data, Recovery of Desulfurization Products,
Applicability to U.S. Coals, and Engineering Design and Cost Estimation.

II. CHEMISTRY AND PROCESS DATA

A. Selection of Specific Reagents and Conditions

1. PROCESS DESCRIPTION

In the Meyers Process (1), aqueous ferric sulfate or chloride selectively
oxidizes the pyritic sulfur content of coal to form elemental sulfur and iron
sulfate. Iron sulfate dissolves in the aqueous solution. The free sulfur may
then be removed from the coal matrix by steam or vacuum vaporization or
solvent extraction, and the oxidizing agent may be regenerated and recycled
(Eqs. 1-4).

$$2Fe^{+3} + FeS_2 \rightarrow 3Fe^{+2} + 2S \tag{1}$$

$$14Fe^{+3} + 8H_2O + FeS_2 \rightarrow 15Fe^{+2} + 2SO_4^{-2} + 16H^+ \tag{2}$$

$$S \cdot COAL \rightarrow S + COAL \tag{3}$$

$$3Fe^{+2} + 3/2[O] \rightarrow 3Fe^{+3} + 3/2[O^{-2}] \tag{4}$$

The aqueous extract solution, which contains iron in both the ferrous
and ferric state, may be regenerated by air or oxygen oxidation of the ferrous

ion to ferric ion (9-11). An advantage of the system is that iron is used to remove iron; thus on regeneration it is not necessary to separate the iron which is extracted from the coal from the metal oxidizing agent. Eventually, accumulated iron can be removed by treating a part of the leach solution (Section III.B.).

Experimentally the coal is treated with aqueous ferric chloride or sulfate solution at approximately $100^{\circ}C$ to convert the pyritic sulfur to elemental sulfur and iron sulfate. Ferric nitrate was not made a major part of this study because of expected nitration and oxidation of the coal matrix by the reactive nitrate function. The aqueous solution is separated from the coal and the coal is washed to remove residual ferric salt. The elemental sulfur which is dispersed in the coal matrix is then removed by distillation or by extraction with a solvent such as benzene, toluene or naptha. The resulting coal is basically pyrite free and may be used as low sulfur fuel.

2. SELECTION OF SPECIFIC REAGENTS AND CONDITIONS

Four coals were selected for initial experimental evaluation (12,13), Pittsburgh, Lower Kittanning, Illinois No. 5 and Herrin No. 6. Their sulfur form distribution is typical of coals east of the Mississippi River and they are representative of major U.S. coal beds.

Analyses of the four coal samples that were used for the study are shown in Table 5-1. These coals were run-of-mine uncleaned samples.

Both ferric chloride and ferric sulfate were evaluated for removal of pyritic sulfur with good results. However, ferric sulfate has the following advantages: a) it is less corrosive to metal reaction vessels, b) regeneration is less complicated and expensive because the $FeSO_4$ formed (Eq. 2) does not have to be separated from iron chloride, c) small amounts of residual leach solution will not contaminate the coal, and d) ferric sulfate appears to be more selective toward pyrite than ferric chloride.

As indicated by the data in Table 5-2, the extent of the reactions indicated by Eqs. 1 and 2, or the molar sulfate-to-sulfur ratio, is 2.4 ± 0.2 when rock pyrite is used but is 1.4 ± 0.4 for sedimentary pyrite found in the coals used in this experimentation and indeed for all coals subsequently investigated. There is always a slight excess reaction of Fe^{+3} with the coal matrix, which would generate excess acid and Fe^{+2} over that formed in a comparable reaction with mineral pyrite. The excess H^+ and Fe^{+2} could have the effect of lowering the oxidizing potential of the solution such that more sulfur survives. However, for coal there were no significant variations in the $SO_4^=/S$ ratio with ferric ion concentration, acid concentration, coal type or reaction time.

TABLE 5-1

Dry Analyses of Coals

Analysis	Lower Kittanning	Illinois No. 5	Pittsburgh	Herrin No. 6
Pyritic sulfur	3.58 ±0.08	1.57 ±0.03	1.20 ±0.07	1.65 ±0.04
Sulfate sulfur	0.04 ±0.01	0.05 ±0.01	0.01 ±0.01	0.05 ±0.01
Organic sulfur	0.67 ±0.10	1.86 ±0.04	0.68 ±0.16	2.10 ±0.06
Total sulfur	4.29 ±0.06	3.48 ±0.03	1.88 ±0.07	3.80 ±0.04
Ash	20.77 ±0.59	10.96 ±0.26	22.73 ±0.48	10.31 ±0.28
Btu	12,140 ±55	12,801 ±58	11,493 ±60	12,684 ±55
Rank	medium volatile bituminous	high volatile B bituminous	high volatile A bituminous	high volatile B bituminous

TABLE 5-2

Sulfate-to-Sulfur Ratio for Extraction of Coal and Mineral
Pyrite with Ferric Solution

Substrate	Sulfate-to-sulfur ratio*,**
Mineral Pyrite	2.4 ±0.2
Lower Kittanning	1.4 ±0.3
Illinois No. 5	1.6 ±0.4
Pittsburgh	1.3 ±0.3
Herrin No. 6	1.4 ±0.3

*Average of all runs ± standard deviation.

**Coal 100 grams of the desired mesh (14 x 0 or 100 x 0), was treated with 600 ml ferric chloride or ferric sulfate, 1M in ferric ion. The solution was brought to reflux ($102^{\circ}C$) for the desired time (usually 2 hours), filtered, and washed thoroughly on the filter funnel. The coal was refluxed with 400 ml toluene for 1 hour to remove the sulfur from the coal, and then the coal was dried at $150^{\circ}C$ under vacuum.

The effects of acid concentration, coal particle size, ferrous and sulfate ion concentrations, and reaction time of pyrite removal were investigated in detail (12). These parameters were studied under conditions that gave 40-70% pyritic sulfur removal so that the effects of parameter variations were clear and not so small as to be masked by experimental error. In addition, studies were performed to demonstrate 90-100% pyritic sulfur removal using both ferric chloride and sulfate, and a set of experiments was performed to illustrate the difference between ferric sulfate and ferric chloride.

The effect of added hydrochloric acid concentration (Table 5-3) was studied in order to define the effect of acid on pyrite and ash removal, sulfate-to-sulfur ratio, final heat content, and possible chlorination of the coal. Acid was studied at 0.0 and 1.2M hydrochloric acid in 0.9M ferric chloride. Duplicate runs were made at each concentration with all four coals for a total of 32 runs. The results showed no definite trends (except one *vide infra*) even when the data were smoothed via computer regression analysis.

TABLE 5-3

Variation of Ferric Ion Consumption with Acid Concentration
and Ferric Anion

Substrate	FE(II)(Exptl)/Fe(II)(Calcd)		
	0.9N FeCl$_3$		0.4N Fe$_2$(SO$_4$)$_3$, 0.0M H$_2$SO$_4$
	0.0M HCl	1.2M HCl	
Lower Kittanning	1.2	1.4	1.2
Illinois No. 5	3.8	6.6	1.6
Pittsburgh	2.2	3.4	1.5
Herrin No. 6	3.7	6.4	2.4

An important consideration in any chemical process is the selectivity of the desired reaction. In the oxidative leaching of pyrite by Fe^{+3}, the extent of the reaction of the reagent with the coal matrix has an effect on the process economics. The extent of this reaction varies from small to substantial depending on the acid concentration, coal, and ferric anion. In order to define this effect quantitatively, the ratio of actual millimoles of ferrous ion produced to the millimoles of ferrous ion necessary to produce the sulfate and elemental sulfur that was recovered was calculated for each run (see Eqs. 1 and 2). This ratio, Fe(II) (experimental)/Fe(II) (calculated), has a value of one for 100% selectivity and a higher value for less than 100% selectivity. The data for ferric chloride (Table 5-3) were smoothed by linear regression analysis using the values generated in the acid matrix; the ferric sulfate values are the average of triplicate runs.

The higher ranked Appalachian coals (Lower Kittanning and Pittsburgh) react to a lesser extent with ferric ion under all experimental conditions than do the lower ranked eastern interior coals (Illinois No. 5 and Herrin No. 6). In addition, the ferric chloride runs show that an acid-catalyzed reaction occurs in this system which is most evident for the Illinois No. 5 and Herrin No. 6 coals. In these coals, a reduction of about 42% in ferric ion consumption is observed when the starting HCl concentration is reduced from 1.2M to 0.0M.

The corresponding reductions for Pittsburgh and Lower Kittanning coals are 35 and 14%, respectively. When ferric sulfate is used, further reductions in ferric ion consumption ranging to 63% for Illinois No. 5 coal are observed. Thus, ferric sulfate is the preferred form of ferric ion for increased selectivity.

Results in terms of both final sulfur values and pyrite removal are given in Table 5-4 for maximum desulfurization. Note that pyritic removal computed from either sulfur forms analyses or the difference in total sulfur between processed and untreated coal (Eschka analysis) resulted in similar values. The observation of greater than 100% removal, computed from differential total sulfur content, is a result of cumulative error in analysis and the removal of small amounts of sulfate (0.02-0.04%).

TABLE 5-4

Pyritic Sulfur Removal Data at $100^{\circ}C$*

| Coal | Total Sulfur Analysis | | | |
	Start (%)	Finish (%)	Total S** removal (%)	Pyritic S removal (%)
Lower Kittanning	4.32	0.93	78	95
Pittsburgh	1.88	0.75	60	95
Illinois No. 5	3.48	1.88	46	102
Herrin No. 6	3.80	2.04	46	107

| Pyritic Sulfur Analysis*** | | |
Start (%)	Finish (%)	Pyritic S** removal (%)
3.58	0.06	98
1.20	0.09	93
1.57	0.10	94
1.65	0.05	97

*Six one-hour leaches with fresh 1M $FeCl_3$ (0.1M HCl).

**Assuming total sulfur removal $= \dfrac{S_o - S_i}{S_o} \times 100$,

where S_o = % sulfur content at start and
S_i = % sulfur content after extraction.

***Based on sulfur forms analysis.

It became apparent that the nominal laboratory washing procedure left a variable (sometimes considerable) amount of chloride on the coal and that the Illinois No. 5 and Herrin No. 6 coals retained more chloride than the Pittsburgh or Lower Kittanning coals. A random sampling analysis showed that the residual chloride level ranged from 0.61 to 3.06% and that further very extensive washing reduced this level to a still environmentally unacceptable 0.40-1.40% (12). Irreversible chloride retention by the coal may be caused by adsorption, formation of insoluble salts, or the relatively nucleophilic Cl⁻ may react with carbonyl and other oxygenated-, nitrogen- or sulfur-containing functional groups in the coal organic matrix. As indicated by the rate data shown in Table 5-5, chloride is reasonably nucleophilic and further, is a much better nucleophile in aqueous solution than $SO_4^=$ (14) by several orders of magnitude.

TABLE 5-5

Rates of Reaction of Nucleophilic Anions with p-Nitrophenyl
Acetate in Aqueous Solution at 25.0° (14)

Anion	K_2 1 mole^{-1} min^{-1}
CH⁻	890
Cl⁻	16
$CO_3^=$	1.06
H_2O	6×10^{-7}

The data in Table 5-6 indicate the relationship of excess ferric ion consumption to dry mineral matter and pyrite free heat content. Note that for the coal samples utilized, there was essentially no measurable heat loss when ferric sulfate was used as the leachant. Later data shows that ferric sulfate reduces the heat content of some coals.

For a 14,000 Btu/lb, dry mineral matter-free coal and 2-3% pyritic sulfur, each 100% excess ferric ion consumption results in a 60-100 Btu/lb reduction in the heating value of the coal. In coals where there are large amounts of pyritic sulfur and minor reaction of Fe^{+3} with the coal matrix, there is an increase in heat content when it is calculated on a dry basis, because of ash reduction caused by pyrite removal. More details are presented in Section IV.B.

TABLE 5-6

Variation of Final Heat Content with Reaction Conditions
Dry Mineral Matter and Pyrite-Free Heat Content* (13)

		Heat Content, Btu/lb		
		Final		
		0.9N FeCl$_3$		0.4N Fe$_2$(SO4)$_3$, 0.0M H$_2$SO$_4$
Coal	Initial	0.0M HCl	1.2M HCl	
Lower Kittanning	15,069	14,963	14,876	15,192
Illinois No. 5	14,278	13,981	13,672	14,020
Pittsburgh	14,787	14,648	14,466	14,742
Herrin No. 6	14,039	13,805	13,506	14,038

*Values are correct for both ash and heat content associated with the exact amount of pyrite present.

A test matrix was performed (summarized in Table 5-7) to compare over-all abilities of ferric sulfate and chloride to remove pyritic sulfur from all four coals. Slightly less sulfur was removed by 0.4N ferric sulfate than was indicated with 0.4N ferric chloride. However, 0.9N ferric sulfate removed an equal or greater amount of sulfur than 0.9N ferric chloride.

Thus, ferric sulfate removes equivalent amounts of pyrite and has several advantages over ferric chloride for use in a desulfurization process.

TABLE 5-7

Comparison of Ferric Sulfate and Chloride for Pyrite Removal* (12)

| | Pyritic Sulfur Removal, wt % | | | |
| | 0.4N Fe^{+3} | | 0.9N Fe^{+3} | |
Coal	Cl	SO$_4$	Cl	SO$_4$
Lower Kittanning	43	38	43	54
Illinois No. 5	48	43	50	50
Pittsburgh	50	33	58	—
Herrin No. 6	35	33	52	64

*Conditions are the same as shown in Table 5-4 except that 0.4N and 0.9N Fe^{+3} solutions were utilized.

B. Rate of Pyritic Sulfur Removal

The preceding studies defined the ferric reagent, solution acidity and general reaction conditions under which Fe^{+3} efficiently leached the pyritic sulfur content of four representative run-of-mine U.S. coals. Process development studies (13) next concentrated on the generation of process design data on a single coal from the Lower Kittanning seam for detailed design of a pilot or process development unit, as well as for the conceptual design and cost estimation for full-scale plants. Attention was focused on the lower Kittanning coal, since it is representative of the Appalachian Coal Basin which is both the region of the highest U.S. coal production and highest process applicability.

Emphasis was initially placed on the processing of water suspendable fine coal (14 to 100 mesh, 1.41 mm to 149μ, top-size). More than 60 fully material balanced leaching experiments were performed for the purpose of obtaining the above data and simulating the process unit operations of leaching, elemental sulfur removal by solvent extraction and washing of residual sulfate. Material balances were obtained for coal, water, iron and iron forms, elemental sulfur, sulfate, and toluene for the Lower Kittanning seam coal and some data was also obtained on an Illinois No. 5 coal. The effect of the following process parameters on the leach reaction rate was also determined: mode of leacher operation, residence time, pyrite concentration, total solution iron and Fe^{+2} concentration temperature, and coal particle size.

Total iron in solution (in the 2.5-12% w/w range investigated) had no measurable effect on the rate of removal of pyritic sulfur from the two coals studied. However, Fe^{+2} concentration, as expressed in the ratio of Fe^{+3} to total iron in solution, and pyrite concentration in coal had a strong effect. It was found that the pyritic sulfur leaching rate could be represented by Eq. 5.

$$r_L = - \frac{dW_P}{dt} = K_L W_P^A Y^B \qquad . \qquad (5)$$

where

r_L is the pyritic sulfur leaching rate, expressed in weight of pyrite removed per 100 weights of coal per hour

W_p is the pyrite concentration in coal at time t in weight percent

t is reaction (leaching) time in hours

K_L is the reaction rate constant, in $(hours)^{-1}$ $(wt\% \text{ pyrite in coal})^{1-A}$

Y   is the ferric ion to total iron ratio in the reactor (leacher) at time t, dimensionless

and

A,B   are reaction orders with respect to W_p and Y, respectively.

The best computer fit of data (K_L value closest to "universal" constancy) for the Lower Kittanning coal was obtained with values of A = 2 and B = 2. The derived empirical rate expression for the removal of pyritic sulfur from Lower Kittanning coal by leaching with Fe^{+3} is shown in Eq. 6:

$$r_L = \frac{dW_p}{dt} = K_L W_p^2 Y^2 \tag{6}$$

where K_L is a function of temperature and coal particle size. The rate constant value for 100 mesh top size Lower Kittanning coal extracted with ferric sulfate solution containing 5 ± 2 wt% total iron, at $102^\circ C$, is

$$0.12 \leq K_L \leq 0.15 (hours)^{-1} (wt\% \text{ pyrite in coal})^{-1}$$

(See Chapter 3-Ib, where the rate of mineral pyrite oxidation by Fe^{+3} was found to be first order in the specific surface area of pyrite and first order in the ratio of $[Fe^{+3}]$ to $[Fe_T]$.)

The surface area of pyrite in coal is essentially unknown but probably highly variable (Chapter 2) so that correlation of the reaction rates with this variable was not attempted.

The empirical rate equation and the computed value of the reaction constant served as the basis for the generation of the process design curves in Figure 5-1. These curves comprise fundamental information for plant design of the leacher section for treatment of suspendable coal.

The curves shown in Figure 5-2 demonstrate the effect of temperature on pyritic sulfur removal. The data indicates that the rate doubles between $70^\circ C$ and $102^\circ C$ and that it may double between $102^\circ C$ and $130^\circ C$; that is, for the same percent of pyrite removal, the $70^\circ C$ extraction time is twice that of the $102^\circ C$ extraction time. This data and additional experimentation indicate that the rate constant is of the form shown in Eq. 7.

$$K_L = A_L \exp(-E_L/RT) \tag{7}$$

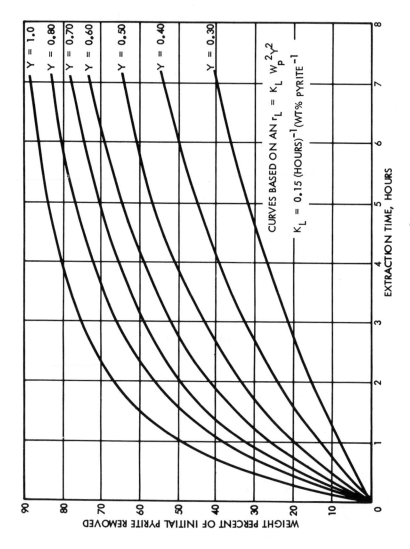

Figure 5-1. Rate of Pyrite Removal from Lower Kittanning Coal as a Function of $[Fe^{+3}]/[Fe_T]$ (13)

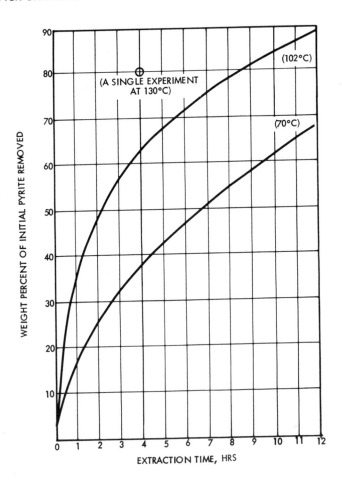

Figure 5-2. Effect of Temperature on Pyritic Removal (13)

The pyritic sulfur removal data from 21 run-of-mine United States coals (representative of major U.S. bituminous coal reserves) is plotted in Figure 5-3. These coals were extracted with ferric sulfate, under conditions previously described, while maintaining $Y > 0.80$. The median pyritic sulfur removal for the 20 coals was 53% in 0.5 hr, 68% in 1.0 hr, 78% in 3 hrs, 87% in 6 hrs, and 90-95% in 23 hrs. The median pyritic sulfur removal data for the 20 coals showed a removal rate in excess of the $Y = 1.0$ rate design curve obtained for the run-of-mine Lower Kittanning coal, *vide supra* (Figure 5-1), though the Y values for the 21 coals were <1.0. Thus, the Lower Kittanning design data base described above is somewhat conservative for most U.S. coals in terms of required leach time.

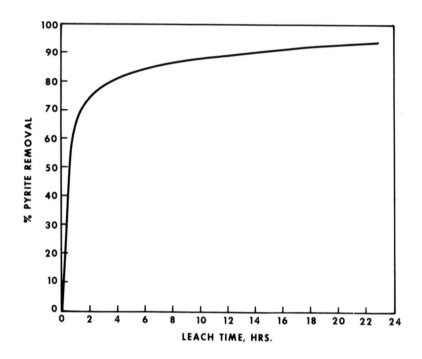

Figure 5-3. Median Pyrite Removal as a Function of Time – 21 U.S. Coals

Hamersma calculated the relative rate constants for each coal as compared with the Lower Kittanning seam coal from its Martinka mine (Table 5-8). The Martinka mine is the coal utilized as the substrate for engineering design studies. This mine and the others from the Lower Kittanning seam are among the slower reacting coals.

It can be seen from Table 5-8 that there is a wide band of rate constants rather evenly spread over a factor of approximately 30. The Kopperston No. 2 and Harris Nos. 1 and 2 coals react more rapidly than the slowest coals (Dean and Muskingum) by a factor of about 30. Thus, it is apparent that real and significant rate differences do exist between pyrite in various coals. Characteristics of coal such as pore structure, size and shape distribution of pyrite, etc., may be the primary factors affecting the rate constant as reflected in the observed band of values found for the rates given in Table 5-8.

The effect of coal top-size was investigated for run-of-mine Lower Kittanning coal (Figure 5-4). Treatment of 1/4 x 0 in. coal particles resulted in removal of 37% of the pyritic sulfur in 4 hrs, 73% in 24 hrs and 81% in 48 hrs. The relative rates of removal for 1/4 x 0 in. vs 100 mesh top sizes are 0.25 at 4 hrs, 0.18 at 24 hrs and 0.12 at 48 hrs.

TABLE 5-8

Relative Rate Constants for Pyritic Sulfur Removal

Coal Mine	Seam	Top size, microns	Sp^o*	$t_{80\%}$**, hrs	$\dfrac{1}{Sp^o t_{80\%}}$	Relative*** rate
Kopperston No. 2	Campbell Creek	149	0.47	2.0	1.1	16
Harris Nos. 1&2	Eagle & No. 2 Gas	149	0.49	2.3	0.89	13
Marion	Upper Freeport	149	0.90	3.0	0.37	5.3
Lucas	Middle Kittanning	100	1.42	3.25	0.22	3.1
Shoemaker	Pittsburgh	149	2.19	2.9	0.16	2.3
Williams	Pittsburgh	100	2.23	3.0	0.15	2.1
Ken	No. 9	149	2.85	2.5	0.14	2.0
North River	Corona	149	1.42	5.0	0.14	2.0
Star	No. 9	100	2.66	3.0	0.13	1.9
Mathies	Pittsburgh	100	1.05	9.0	0.11	1.6
Powhattan No. 4	Pittsburgh No. 8	75	2.75	4.0	0.091	1.3
Homestead	No. 11	149	3.11	3.5	0.092	1.3
Fox	Lower Kittanning	75	3.09	4.5	0.072	1.0
Isabella	Pittsburgh	149	1.07	13.0	0.072	1.0
Martinka	Lower Kittanning	149	1.42	10.0	0.070	1.0
Meigs	Clarion 4A	149	2.19	8.5	0.054	0.77
Bird No. 3	Lower Kittanning	100	2.87	8.0	0.044	0.62
Dean	Dean	100	2.62	10.2	0.037	0.53
Muskingum	Meigs Creek No. 9	149	3.65	8.0	0.034	0.49

*Initial pyritic sulfur concentration in wt%.

**Time to 80% pyrite removal

***$1/Sp^o t_{80\%}$ relative to value for Martinka Mine.

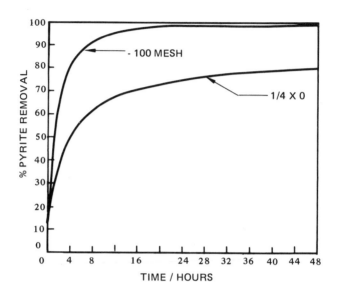

Figure 5-4. Effect of Coal Top-Size on Pyrite Leaching at 100°C

Up to this point, the experimentation reported was concerned with desulfurization of run-of-mine coal. However, a large portion of U.S. coal is cleaned at the mine to remove low Btu rock material (see Chapter 2). A narrow mesh size-fraction (8 x 14 mesh) of the 1/4 x 0 in. top-size Lower Kittanning coal, comprising 25% w/w of the total material (16), was extracted with ferric sulfate under the conditions described above. A rate of removal very similar to that of the entire 1/4 x 0 in. grind was observed. Thus, the 8 x 14 mesh fraction is representative of the reaction rate of the 1/4 in. top size. A portion of the 8 x 14 mesh size fraction was cleaned by float-sink at 1.75 density and the float portion (80% w/w of the sample) was leached with ferric sulfate (Figure 5-5). The results showed that the clean coal reacts substantially faster than run-of-mine coal (80% removal in 24 hrs and >90% in 48 hours) and that coarse coal (in this case a size-gravity fraction) can be chemically desulfurized to near zero pyrite, e.g., 0.09%. These results, if extrapolative to other U.S. coals (as would be anticipated), have important implications for the eventual mode of application of this and possibly other extraction methods.

The data for removal of the final 1% of pyritic sulfur mesh coal is replotted in Figure 5-6. Here it can be seen that the last 0.5% pyritic sulfur is very slowly removed from run-of-mine coal, while clean coal proceeds to nearly complete removal.

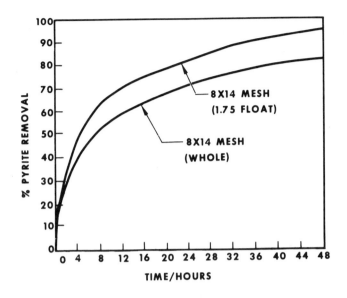

Figure 5-5. Extraction of Clean and Run-of-Mine Coal with Ferric
            Sulfate at $100^\circ C$

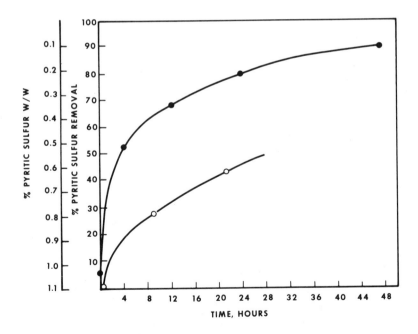

Figure 5-6. Removal of Final 1% of Pyritic Sulfur from Whole and
            Clean Coarse Coal at $100^\circ C$

C. Regeneration of Ferric Sulfate Leach Solution

The chemistry of oxidation of Fe^{+2} to Fe^{+3} by air or oxygen in aqueous solution (17) is well known. A number of processes have been developed for the recovery of sulfuric acid from ferrous sulfate (18) or hydrochloric acid from ferrous chloride (19) which involve the oxygen oxidation of Fe^{+2} to Fe^{+3} as a first step, concurrent with or followed by hydrolysis, depending on temperature and other conditions.

The reaction of ferric sulfate with iron pyrite in coal produces ferrous sulfate, sulfuric acid and elemental sulfur in the molar amounts as shown in Eq. 8 for the observed SO_4^{-2}/S ratio of 1.5.

$$FeS_2 + 4.6 \ Fe_2(SO_4)_3 + 4.8 \ H_2O \ \rightarrow \ 10.2 \ FeSO_4 + 4.8 \ H_2SO_4 + 0.8S \quad (8)$$

Only a portion of the ferrous sulfate must be regenerated to maintain the acid at a constant level (Eq. 9),

$$9.6 \ FeSO_4 + 4.8 \ H_2SO_4 + 2.4 \ O_2 \ \rightarrow \ 4.8 \ Fe \ (SO_4) + 4.8 \ H_2O \quad (9)$$

and even less ferrous sulfate must be regenerated to provide the exact amount (4.6 moles) of ferric sulfate for recycle. The rejection of these process products are discussed in Section III.B.

Regeneration studies (13) were performed for depleted leach solution at temperatures below the point at which irreversible hydrolysis occurs and without the presence of coal. The regeneration rates were obtained under conditions where oxygen in the form of minute air or oxygen bubbles was dispersed throughout the iron sulfate solution. Thus, an attempt was made to continually saturate the solution with oxygen at the partial pressure of oxygen present in the regeneration gas. Minute bubbles were formed by pumping a portion of the liquid in turbulent flow ($N_{Re} > 3000$) through a pipe whose length was 50 or more times its diameter. Gas containing oxygen was added to the liquid in an amount ranging from less than 1% to greater than 10% volume gas/volume liquid at turbulent flow conditions. This method is very similar to aeration equipment used to reduce the biological or chemical oxygen demand of chemical plant effluent streams, except that ferric sulfate regeneration is conducted at higher temperatures and pressures.

Extensive experimentation was performed while varying the following parameters: air vs oxygen, pressure (30-150 psig), temperature (70-160°C), liquid flow rate (1.6-7.6 l/min), gas flow rate (0.25-2.3 l/min), total iron (3.3-10% w/w), starting Fe^{+2}/Fe ratio (0.15-1.0), final Fe^{+2}/Fe (0.23-0.009), gas/liquid mixing (residence time and velocity) and reactor tank solution volume (0.5 to 1.5l).

The regeneration rate was found to be second order in $[Fe^{+2}]$ over the range of Fe^{+2} concentration from 100% to less than 1% w/w of the total iron. The rate is second order in Fe^{+2} and first order in O_2 (Eq. 10) as first reported by Pound (17).

$$r_R = \frac{-d[Fe^{+2}]}{dt} = K_R [Fe^{+2}]^2 [O_2] \qquad (10)$$

where

$[Fe^{+2}]$ = concentration of ferrous ion, mole/liter

$[O_2]$ = oxygen partial pressure, atm

K_R = 1.836 l/mole-atm-hour at 120°C

Over the temperatures range 100°-130°C, the rate constant was found to vary exponentially with temperature (Eq. 11).

$$K_R = 40.2 * 10^6 \exp(-13,200/RT) \qquad (11)$$

The regeneration reaction is quite exothermic, since $\Delta H = -18.6$ Kcal/g-mole of ferrous sulfate oxidized.

D. Simultaneous Pyrite Leaching and Ferric Sulfate Regeneration

1. SELECTION OF CONDITIONS

Process plant design for coal feeds with (a) low pyrite sulfur concentration (<1% w/w) or (b) leach rates grossly higher than that found for Lower Kittanning coal could reach a capital cost minimum utilizing separate leach and regeneration. Also the processing of coarse, essentially non-water suspendable coal (e.g., 1/4 in. top size) would probably be more economical using separate leach and regeneration. However, combining the leach and regeneration steps in the chemical desulfurization of fine, suspendable coal (e.g., 14 mesh top size) with ferric sulfate is thought to be economically attractive for coals which have a demand for (a) relatively large amounts of Fe^{+3} and/or (b) maintenance of a high Y value (see Section II.B). Combining the two steps in one reactor could provide a net capital cost savings, provided that both the leach and regeneration reactions operated efficiently under identical conditions of temperature, pressure and mixing.

Initial experimentation was reported (Table 5-9) in which oxygen was introduced through a glass frit into a ferric sulfate-coal slurry. The results showed (a) a small amount of regeneration could be obtained at 95°C utilizing this method for dispersal of oxygen into the solution and (b) $CuSO_4$ catalyst and lower temperature increased the degree of regeneration (13). This is in concert with reports by other authors that copper sulfate approximately doubles the rate of ferrous sulfate oxidation (20,21).

TABLE 5-9

Initial Simultaneous Ferric Sulfate Leach and Regeneration Experimentation*

Condition	Pyritic sulfur removal %/w/w	% Regeneration of Fe^{+3}
N_2 flow/95°C	69	0
O_2 flow/95°C	70	15
O_2 flow with catalyst**/95°C	66	19
O_2 flow with catalyst**/65°C	68	60

*100g, Illinois No. 5 coal treated with 2.4 liters of 1.0 N ferric sulfate for 4 hrs with O_2 flow: 200 ml/min through fine fritted glass at bottom of flask.
**0.05M in cupric sulfate.

It is significant to note that the presence of oxygen had no effect on the degree of removal of pyrite, i.e., there was no separate competing direct reaction of oxygen with pyrite. Further, due to the large excess of Fe^{+3} present in the reaction vessel, Fe^{+3} regeneration did not increase the Fe^{+3}/Fe_T ratio enough to cause a noticeable increase in rate of pyrite leaching. This method of simultaneous leach and regeneration was not pursued, however, because utilization of oxygen on a once-through atmospheric-pressure basis would be inefficient. Maier (21) reports that the efficiency of utilization of oxygen for oxidation of ferrous sulfate utilizing a static atmospheric aeration technique is only 0.0004%. Further, since the rate of regeneration of ferric sulfate solution depends on $[Fe^{+2}]^2$, it was clear that the regeneration (which was obtained at high $[Fe^{+2}]$) would fall off rapidly as regeneration proceeds and as the Y = 0.8-0.9 region needed for rapid leaching is approached.

2. INITIAL ENGINEERING DATA

Both the leaching (Section B.) and regeneration (Section C.) rate studies indicated that temperatures of 100-130°C are favorable for rapid reaction. The regeneration studies indicated that regeneration was a function of oxygen pressure and that 4-6 atm of oxygen would provide a rapid rate of regeneration. tion. Further, there is no measurable oxidation of Appalachian Basin coal by oxygen up to 6 atm and 120°C (15).

Therefore, since the leach and regeneration reactions seemed entirely compatible the combination of the two processes was investigated under the above conditions (13). It was necessary to use an apparatus which would provide intimate contact (fine bubbles) of oxygen with the leach solution slurry. Static mixing, gas dispersion by impellor, or turbulent flow methods were all suitable for this task. Since an apparatus was available for injection of oxygen into the discharge line of a slurry pump (turbulent flow method), this method was utilized. The following variables were investigated: coal top size (14 and 100 mesh), slurry solids (20 and 33% w/w), total solution iron content (from no added iron to 7% w/w in solution), temperature (110-130°C), and pressure (60-85 psi of oxygen, 100 psig maximum including partial pressure of water).

The design of a simultaneous leach and regeneration process scheme involves three pyrite removal operations (Table 5-10). First, dry coal (Lower Kittanning) and leach solution are mixed and heated to reaction temperature (Operation 1), while roughly 20 wt% of the pyrite in the coal reacts with the ferric sulfate in the leach solution. Typically, the entering leach solution is partially depleted (low Y), having been recycled from the third process leach operation. Secondly, simultaneous leach and regeneration (Operation 2) is effected by contacting slurry with oxygen in a pressurized reactor for a period of 1-5 hrs, simultaneously increasing the pyrite conversion to 80-90% and increasing the Y value. Combining Eqs. 8 and 9 for separate leach and regeneration, Eq. 12 is obtained for the combined system where

$$FeS_2 + 2.4O_2 \rightarrow 0.2\ Fe_2(SO_4)_3 + 0.6\ FeSO_4 + 0.8S \tag{12}$$

TABLE 5-10

Simultaneous Ferric Sulfate Leach and Regeneration
Based Pyrite Removal Process

Operation	Temp °C	Pyrite removal	Y
1. Slurry mixing and and heating	20-120	20	0.8-0.5
2. Simultaneous leaching and regeneration	120	80-90	0.5-0.9
3. Settling	90	90-95	0.9-0.8

The SO_4^{-2}/S ratio remains at 1.5, and solution neutrality is maintained. Lastly, the slurry is transferred to a secondary reactor where it is allowed to both settle (concentrate solids) and continue to react at atmospheric pressure (Operation 3) to the 90-95% level. The simultaneous leach and regeneration step is not used to bring the leach reaction to the 90-95% level since economic studies indicate that atmospheric pressure residence time obtained while the slurry is thickened is somewhat advantageous to additional residence time in a pressurized reactor. The clarified leach solution is then decanted from the slurry and the coal is filtered from the solution and treated for removal of generated elemental sulfur, while the leach solution is recycled to the first operation.

The experimental results from an illustrative (16) simultaneous leach and regeneration run are shown in Figure 5-7, where simultaneous leach and regeneration begins with a low Y leach solution and a coal fraction which is 20% reacted. In approximately 90 minutes, the coal pyrite is 85% reacted and the leach solution is regenerated to the 0.9 Y level.

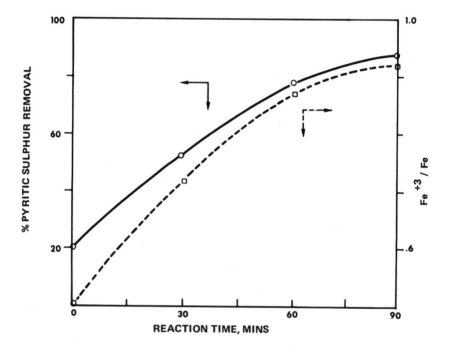

Figure 5-7. Simultaneous Leach and Regeneration at 120°C/85 psig of O_2

It was found (13) that the rate of pyrite removal was not dependent on the total iron content of the solution from a range of "no added iron" to 5% w/w. It was found that the slurry could be used at a high solids content, up to 33% w/w, without affecting the rate. Further, within the temperature range (110-130°C) and pressure range (60-85 psig of oxygen), no experimental differences were found in pyrite removal or iron solution regeneration rates. The ranges of pressure, temperature and solution iron whiich are considered applicable for the simultaneous leach and regeneration system are shown in Figure 5-8. The combination of high oxygen pressure and temperature (upper front volume element of Figure 5-8) is unfavorable for processing, since the coal tends to combust in this region.

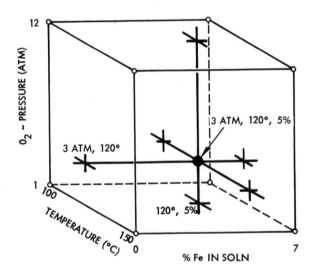

Figure 5-8. Simultaneous Leach and Regeneration – Temperature, Pressure and Solution Iron Concentration

Engineering results appear to be favorable in the region centering at 3 atm, 120°C and 5 wt % iron. Elemental sulfur was isolated from the leached coal in amounts (SO_4^{-2}/S ratios of 1.5-2.0) similar to that obtained for separate leach of pyritic sulfur. Thus, the elemental sulfur formed is not significantly oxidized by oxygen under the conditions utilized. This is not surprising, since non-vapor phase sulfur is not readily attacked by oxygen. It is known that the ignition temperature of sulfur in air is approximately 266°C. One atmosphere of pure oxygen has no noticeable effect in the ignition temperature and at 252°C the rate of sulfur oxidation is roughly a first order function of oxygen pressure (22). Further, elemental sulfur has been shown to be stable at $110^\circ C/150$ psi O_2 for at least 5 hrs during investigation of a pyrrhotite aqueous leaching process (23). It would appear that the mechanism of pyrite oxidation (even in the presence of large amounts of oxygen) continues to involve Fe^{+3} as the primary reactant or chain carrier, since the rates of reaction and sulfur formation during simultaneous leach and regeneration and separate leaching with ferric sulfate are essentially identical. A similar identification of Fe^{+3} as an intermediate in pyrite oxidation in aerated solution was advanced by Singer and Stumm for the mechanism of acid mine drainage (24).

As noted above, there was no effect of solution iron concentration on the reaction rate. Investigations were performed utilizing pure water in place of leach solution in Operations 1 and 2 (Figure 5-9). In these experiments the natural iron sulfate content of the Lower Kittanning coal (0.24% w/w sulfur as sulfate) dissolved during mixing, probably remaining largely in the pores of the coal giving a localized high solution iron concentration. On treatment in the simultaneous leach/regeneration system (Operation 2), more iron was leached from the coal, giving a rapid build-up to 0.5-0.7% w/w total solution iron. Overall, pyritic sulfur removal is somewhat less in 1-1/2 hr reaction period (Figure 5-9) and because of the relatively low solution iron content, very little additional pyrite can be removed in the settler (Operation 3), although much longer residence times in the simultaneous leach regeneration reactor could be expected to give higher removals.

The SO_4^{-2}/S ratios were 1.5-2, indicating the same or a very similar mechanism of Fe^{+3} leaching of pyrite obtains even in low $[Fe^{+3}]$ solutions. Obviously, recycle of the spent leach solution would continue to build up iron sulfate content quickly to the 3-7 wt% iron level range which is normally used in the simultaneous leach and regeneration system unless the leach solution was continually purified by removing the generated iron sulfate. This could be accomplished by treatment of the dilute solutions with lime, but would probably be expensive relative to simply allowing the iron to build up to a level where crystallization or liming of a concentrated iron solution could be performed (See Section III.B.).

Additional runs were made with water acidified with sulfuric acid and with water made alkaline with sodium hydroxide to determine the effect of pH on reaction rate and on elemental sulfur make. No effect was found on either rate or elemental sulfur generation (See Chapter 3, Section I.B.2.c.).

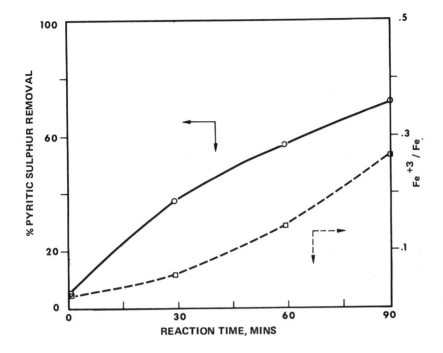

Figure 5-9. Simultaneous Leach and Regeneration at $120^{o}C/85$ psig of O_2 – "No Added Iron"

3. ADVANCED ENGINEERING DATA

a. Major Findings

A detailed bench-scale laboratory assessment of the Meyers Process utilizing a run-of-mine Lower Kittanning seam coal was performed by Koutsoukos et al (16). The results were presented in a government report which is liberally quoted and paraphrased here.

Koutsoukos reaffirmed the rate expressions for leaching and regeneration developed in his earlier work (13). He found that pyrite removal rate in the mixer (operation 1 of Table 5-10), L-R reactor (operation 2), and ambient pressure reactor (operation 3) is governed by the empirical leaching rate expression, Eq. 13.

$$r_L = -\frac{dW_p}{dt} = K_L W_p^2 Y^2 \tag{13}$$

where

W_p = wt percent pyrite in coal

Y = ferric ion-to-total iron ratio in the reactor reagent, and

K_L = rate constant, a function of temperature and coal top size.

Up to approximately 80 percent pyrite removal, the rate constant shows a strong temperature dependence expressible by Eq. 14.

$$K_L = A_L \exp(-E_L/RT) \tag{14}$$

Beyond 80 percent removal, the K_L value appears virtually unaffected by temperature in the range investigated ($90^\circ C$ to $130^\circ C$). Apparently, a change of reaction mechanism occurs when pyrite removal exceeds 80 percent; it is speculated that a diffusion controlled process takes over. This change in mechanism may be specific to the particular high-ash coal utilized by Koutsoukos *et al.*

The E_L value is independent of temperature, coal top size (up to 14 mesh) or reactor unit operation (processing conditions). The value of A_L is independent of temperature under mixer and L-R operations, but different for each operation. The reasons for this unexpected A_L value dependence on mode of operation were not apparent from the available data.

Reagent regeneration is governed by the rate expression, Eq. 15.

$$r_R = -\frac{dFe^{+2}}{dt} = K_R P_{O_2} (Fe^{+2})^2 \tag{15}$$

where

K_R = $A_R \exp(-E_R/RT)$

P_{O_2} = oxygen partial pressure, and

Fe^{+2} = ferrous ion concentration in the reagent solution;

A_R and E_R are constants.

The reagent regeneration rate operates simultaneously with the leaching rate in the L-R reactor (operation 2).

Other key findings include: 1) the K_L value obtained with 100 mesh top size L.K. coal was only approximately 20% larger than the K_L obtained with 14 mesh top size coal, but the coal top size effect on K_L becomes substantial with ROM coals coarser than 14 mesh top size, 2) since solid-liquid separation efficiency increases substantially with coal top size, it appears preferable to process 14 mesh coal rather than 100 mesh in view of the small rate penalty involved, 3) the concentration of coal in the slurry and the concentration of iron in the reagent affect the leaching rate through their effect on Y, but these effects become negligible under continuous reagent exchange or L-R processing, 4) thus, in principle, the allowable coal content of the slurry is limited only by equipment ability to transfer thick slurries, and the allowable iron content of the reagent is limited only by solubility, and 5) the sulfate sulfur-to-elemental sulfur ratio of (1.5:1) of the product sulfur of the Meyers Process is independent of the mode of processing or the value of the processing parameters used within the ranges investigated. Note that experimentation on Fe^{+3} leaching of mineral pyrite shows an $SO_4^=/S$ pH dependence (Chapter 3-IB).

b. Detailed Results

Simultaneous leach and regeneration pyrite removal data is summarized in Figure 5-10 (16). Koutsoukos concluded that although slurry solids and O_2 pressure are varied, the only discernible parametric effect is slurry concentration during the first hour of processing. The "outlier samples" show a slower rate of pyrite removal for an observed low Y due to oxygen deficiency during the experimentation. Thus, under efficient regeneration conditions the pyrite removal rates should be at least as high as the upper bounds of the data plotted in the figure. The K_L value at 120°C under L-R processing conditions is five times higher than the K_L value for 102°C processing under ambient pressure conditions (separate leaching and regeneration); both the 14 and 100 mesh top size coals exhibit the same large increase in K_L value. The estimated E_L justifies only a twofold increase in K_L for 20 degree temperature rise-observed between 80° and 102°C (separate leaching-regeneration) and between 110° and 130° (L-R processing). The 250 percent increase in the K_L value, which appears to be the result of changing the mode of processing, could not be attributed to reaction of oxygen with pyrite.

Pyrite concentration and the reciprocal of $[FeS_2]$ in coal plotted against L-R reaction time is shown in Figure 5-11.

Koutsoukos stated that the data in Figure 5-11 indicate that L-R processing can best be represented by Eq. 13 through the use of three K_L values. For the 14 mesh top size L.K. coal processed at 120°C, the K_L values are as follows:

$$K_{L1} = 0.5 \text{ (hours)}^{-1} (\% \text{ pyrite in coal})^{-1} \text{ for } W_p \geq 1.6 \text{ wt.\%}$$

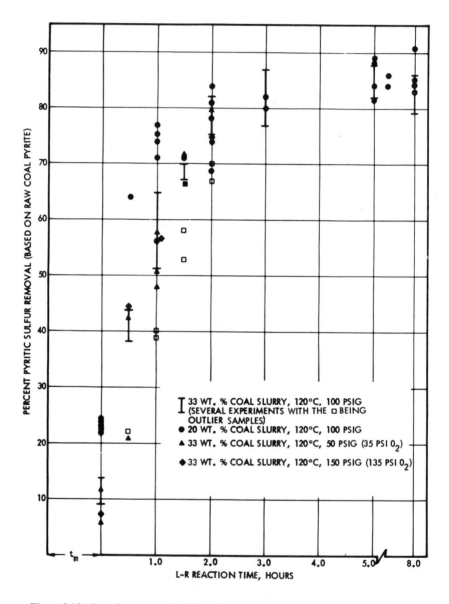

Figure 5-10. Data Summary — Pyritic Sulfur Removal from 14 Mesh Top Size Lower Kittanning Coal Processed at 120°C with 5 wt % Fe Reagent (16)

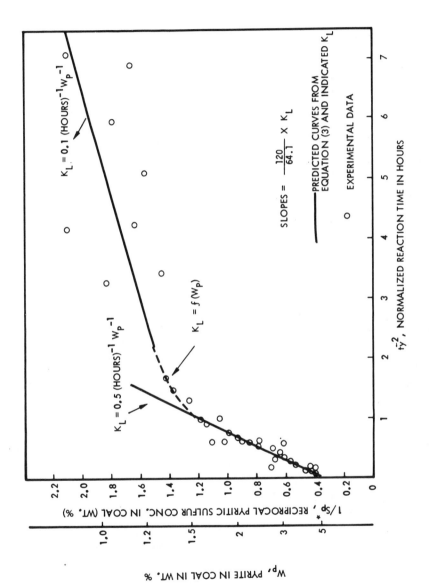

Figure 5-11. Pyrite Leaching Rate Constant Data (16)

$$K_{L2} = f(W_p) = W_p - 1.1 \text{ (hours)}^{-1} \text{ (\% pyrite)}^{-1} \text{ for } 1.6 \geq W_p \geq 1.2 \text{ wt.\%}$$

$$K_{L3} = 0.1 \text{ (hours)}^{-1} \text{ (\% pyrite in coal)}^{-1} \text{ for } W_p \leq 1.2 \text{ wt.\%}$$

The deviation in the K_{L1} is estimated to be ±10%, but the deviation in the K_{L3} value is closer to ±50%. However, the large deviation in K_{L3} is understandable since it can result from only a ±0.1 deviation in the S_p values which comprise the concentration region to which K_{L3} applies; such deviation is normal for pyritic sulfur analyses.

Koutsoukos also noted that the estimated K_{L3} value is identical to the K_L value found applicable for ambient pressure processing of the same size coal at 102°C. Within experimental uncertainty, the same observation was made with the 100 mesh top size coal where K_{L3} was estimated to be 0.12 (hours)$^{-1}$ $(W_p)^{-1}$. The implication of this observation is that if K_{L3} is temperature insensitive during small temperature changes (e.g., a diffusion constant), it is possible to have an observable change in reaction mechanism at 120°C but not at 102°C, provided that at 102°C K_{L1} and K_{L3} are nearly identical.

The temperature effect on pyrite removal during L-R processing of 20% solids 100 mesh and 33% solids 14 mesh top size L.K. coal is shown in Figures 5-12 and 5-13, respectively. Koutsoukos pointed out that the data of Figures 5-12 and 5-13 show that the rate of pyrite removal from suspendable L.K. coal during L-R processing increases with increasing temperature. In addition, the temperature effect on pyrite removal is larger between 110°C and 120°C than between 120°C and 130°C and more pronounced for 33 wt % slurries than on 20 wt % slurries. The 130°C curves of the two slurry concentrations are nearly identical. Typical process design curves for the Meyers Process are shown in Figure 5-14.

Koutsoukos noted that under L-R processing, the parametric effects on the individual rates and process performance differences in the mixer are substantially tempered because of the simultaneous influence of the leaching and regeneration rates on process performance in the L-R reactor. Note that each curve represents both 14 and 100 mesh top size coals; thus, the coal particle size effect virtually disappeared for coal top sizes up to 14 mesh. The slurry coal concentration in the 20-33 wt % range had virtually no effect on overall process performance even though twice as much coal per unit time was processed when 33 wt % slurries were employed as when 20 wt % slurries were used. The effect of pressure (oxygen partial pressure) on overall process performance was also small under L-R processing.

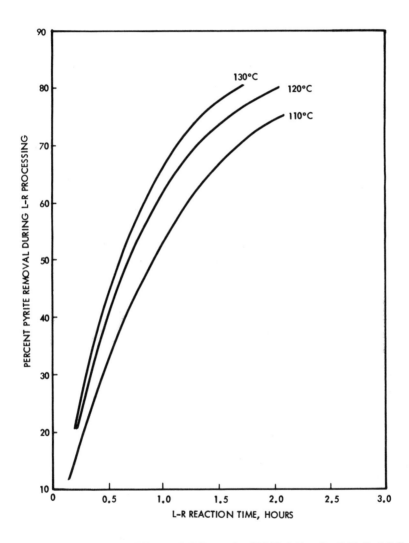

Figure 5-12. Temperature Effect on L-R Processing 100 Mesh Top Size L.K. Coal (16)

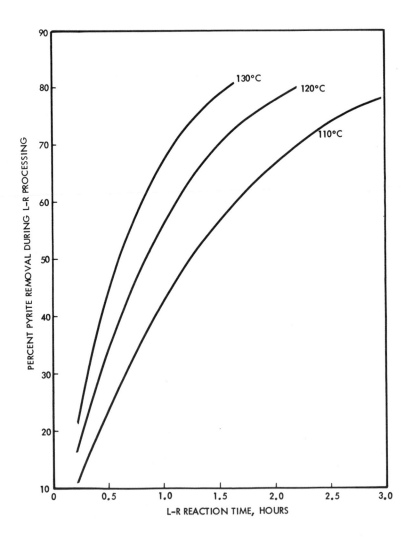

Figure 5-13. Temperature Effect on L-R Processing of 14 Mesh Top Size L.K. Coal (16)

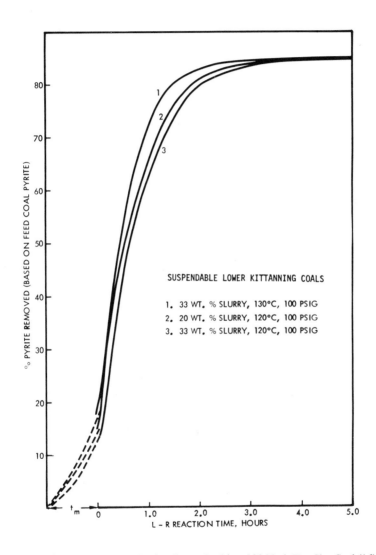

Figure 5-14. Typical Process Design Curves for 14 or 100 Mesh Top Size Coal (16)

Lower Kittanning coal was used for the above investigations. When similar process studies were performed using coal from another seam, the Upper Freeport seam (Marion mine), the leaching rate sharply increased (Figure 5-15).

Koutsoukos noted that appproximately 82% of the pyrite in the Marion coal was removed during the 1.5 hours slurry residence time in the mixer; at this point the pyritic sulfur content of the coal was only 0.16 wt percent. An additional 1.5 hours of L-R processing at 120°C increased the overall pyrite removal

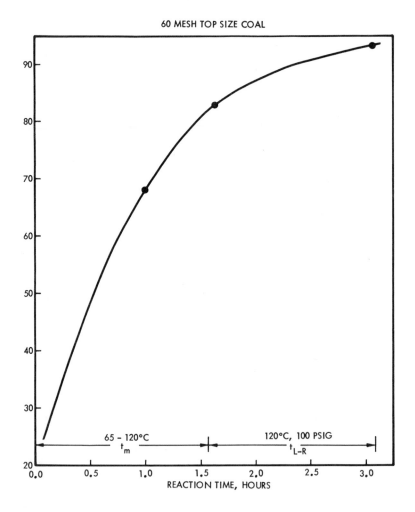

Figure 5-15. Leaching of Marion Mine Coal with 5 wt % Fe Reagent
(60 Mesh Top Size Coal)

to 93 percent and reduced the pyritic sulfur content of the coal to 0.07 wt percent. Obviously, pyrite removal from the Marion coal by the Meyers Process appears to be efficient down to virtually zero pyrite.

Pyrite removal data from nearly identical processing of Marion and L.K. coals are compared in Figure 5-16. It is obvious that pyrite removal from the Marion coal takes place at vastly higher rates than from L.K. coal. Since only a single rate experiment was performed with the Marion coal, it was not possible to determine if pyrite leaching from this coal could be expressed by Eq. 13; the empirical leaching rate expression developed from the L.K. investigations. If, however, it is assumed that Eq. 13 applies to Marion coal, then the data in Figure 5-16 appear to indicate that the K_L value for the Marion coal is between one and two orders of magnitude higher than the equivalent K_L value for the L.K. coal. Note that the two curves in Figure 5-16 are not directly comparable in terms of Eq. 13 because the W_p in the coals differ substantially.

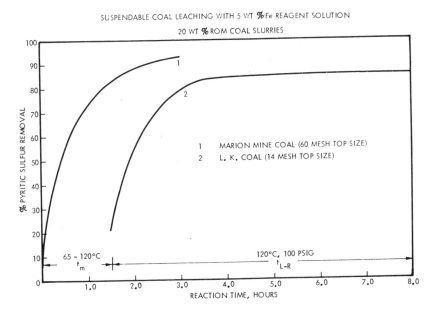

Figure 5-16. Pyrite Leaching – Upper Freeport and Lower Kittanning Coals (16)

III. RECOVERY OF DESULFURIZATION PRODUCTS

Iron sulfates and elemental sulfur are the desulfurization products of the reaction of ferric sulfate with coal. The latter product can be removed from the coal matrix by vaporization, solvent extraction, or chemical reaction and the former by water extraction or pyrolysis. It is also necessary to remove iron sulfates and any dissolved ash components from leach solution prior to recycle in order to maintain a constant solution composition. Methods for removal of elemental sulfur from coal are discussed in the first section below and methods for rejection of iron sulfate and dissolved metal salts from the leach solution and removal from coal are presented in the second section.

A. Removal of Elemental Sulfur from Coal

Two methods for the removal of elemental sulfur from coal, vaporization and solvent extraction have been evaluated and a third, that of chemical reaction, remains to be investigated.

1. VAPORIZATION

Although the atmospheric boiling point of sulfur is 444.6°C (22) which is somewhat above the decomposition temperature of coal, an appreciable vapor pressure (sufficient for engineering design of equipment for removal of sulfur in an inert gas stream) develops at about 250°C (Figure 5-17). At 350°C (which is just below the thermal decomposition point of coal) the vapor pressure of sulfur is nearly 0.25 atmosphere. Therefore, it seems reasonable that elemental sulfur could be removed from coal by vaporization either in an inert gas stream or under vacuum, provided that the sulfur does not extensively react with coal during the thermal process. Mazumdor (3) has studied the reactions of elemental sulfur with coal at temperatures of 400-600°C, wherein sulfur was shown to recombine and also to form hydrogen sulfide by reaction with coal.

The initial demonstration of the removal of pyritic sulfur from coal by ferric oxidation was performed using vaporization for removal of elemental sulfur (2). In this case, vaporization was effected at 100-120°C and reduced pressure provided by a water aspirator with no detectable coal decomposition or elemental sulfur reaction with coal. In fact, greater than 99% pure sulfur was obtained as condensate on a cooled surface (5). Initial attention was focused on vaporizing sulfur from coal at atmospheric pressure and in an inert gas stream (Table 5-11).

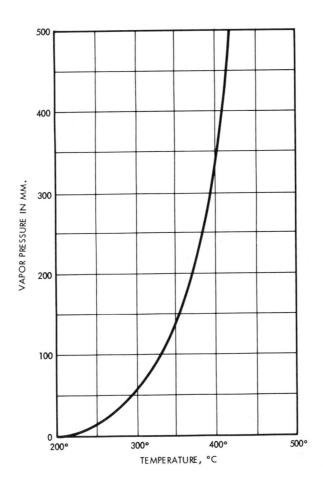

Figure 5-17. Vapor Pressure of Sulfur

A coal sample was treated with ferric sulfate under the usual leaching conditions (16) but was neither water washed to remove iron sulfate from the coal pore structure nor extracted with solvent to remove elemental sulfur (Sample No. 1). Drying was effected at relatively low temperature in an attempt to remove water but not to vaporize large amounts of the elemental sulfur which had been formed. Another sample (No. 2) was washed with water and toluene in order to remove sulfate and elemental sulfur, respectively. The

TABLE 5-11

Vaporization of Elemental Sulfur from Ferric Sulfate Leached Coal*

Sample** No.	Water washed	Toluene extracted	Thermal treatment	% w/w sulfur		% Weight loss during vaporization
				Sulfate	Organic	
1	0	0	0	0.34	1.18	–
2	X	X	0	0.13	0.52	–
3	X	X	220°C/15 mm, Hg	0.05	0.51	2.7
4	X	0	220°C	0.03	0.92	1.7
5	X	0	250°C	0.01	0.63	1.2
6	X	0	350°C	0.01	0.56	2.8

*Lower Kittanning seam coal.

**Coal, 200g 100 x 0 mesh extracted twice with 2.5l 1N $Fe_2(SO_4)_3$ solution, total residence time 22 hrs. Cursory water wash to remove surface iron sulfate; dried at 110°C under vacuum to remove moisture and as little volatile elemental sulfur as possible.

decrease in organic sulfur between Sample Nos. 1 and 2 (0.66% w/w) is a measure of the elemental sulfur which had remained adsorbed on the coal. Another sample (No. 3) was further treated under vacuum at 200°C to remove any additional residual sulfur. There was no further decrease in organic sulfur indicating that a maximum of the elemental sulfur had been removed by the toluene extraction; however, there was a decrease in the sulfate content. This may be due to the thermal decomposition of iron sulfates to iron oxide and sulfur trioxide.

Temperature Programmed Thermogrammetric Analysis studies (26) have shown that iron sulfates quantitatively decompose at 450-600°C in an inert atmosphere giving a residue of iron oxide. Apparently, this degradation occurs at somewhat lower temperature under vacuum and in the presence of a reducing coal environment. Alternatively, the disappearance of sulfate may be accounted for by reaction with sulfur, since iron sulfate has been reported to react with molten elemental sulfur to form sulfur dioxide and iron sulfide. The iron sulfide formed would analyze as organic sulfur (27) thus accounting for the slightly higher organic sulfur measured for sample 6 vs 2. Note that the weight loss during vacuum treatment was only 2.7% of which 0.67% could be ascribed to elemental sulfur.

Three samples were contacted with an inert gas stream (argon) in small ceramic boats inserted into a Burrell tube furnace at the specified temperatures and residence times (Nos. 4-6). It can be seen from Table 5-11 that increasing the temperature of vaporization from 200-350°C gives increasing sulfur vaporization. Thus, a vaporization temperature of 200-350°C is indicated for use in process design. Note that the weight losses during vaporization did not exceed 2.8% (including the loss due to volatilization of sulfur) indicating that sulfur can be relatively selectively vaporized from coal. The method utilized for contacting the coal with the inert gas stream was rather primitive compared to that which could be obtained in commercial size equipment, so that further improvement could be expected.

2. SOLVENT EXTRACTION

In the course of the development of the various unit operations associated with the Meyers Process (13) it was found that a number of solvents can penetrate the coal matrix and dissolve elemental sulfur which has been formed by reaction of ferric ion with pyrite. Filtration then provides a filtrate rich in elemental sulfur. This experimentation is the first report of elemental sulfur having been removed from coal by solvent extraction. The choice of solvent involves the following criteria: a) range of sulfur solubility as a function of temperature so that coal can be extracted with hot solvent and sulfur recovered by crystallization from cool solvent, b) the solvent must be near totally recoverable from coal, c) trace amounts of solvent retained on coal must not provide a source of contamination, and d) the solvent must be available in large quantities for ultimate commercial development.

Two basic solvent classes can be distinguished for sulfur. These are solvents which dissolve the sulfur by physical interaction and those which involve reactive solution in which sulfur during the solvation process forms new chemical species. Examples of the latter type are sulfur trioxide, concentrated sulphuric acid and liquid ammonia. These solvents are relatively unsuitable for direct recovery of sulfur because they tend to react with coal. However, they could provide a basis for secondary recovery of sulfur from a non-reactive solvent — *vide infra*.

Some representative examples of non-reactive solvents are shown in Table 5-12. Additional solvents which are non-reactive with sulfur (at moderate temperatures) include ethylene dibromide, chlorobenzene, carbon tetrachloride, kerosene, acetic acid and the like. Halogenated solvents are undesirable because retention of even small amounts on the coal would cause contamination leading to a halogen air pollution emission problem as well as to equipment corrosion when coal is utilized as fuel.

TABLE 5-12

Solubility of Sulfur in Selected Solvents (28)

Solvent	Solubility, g/100g solvent	Temp °C
Benzene	10.5	80
	0.25	20
Carbon disulfide	50	40
	29.5	20
Coal tar oil	27	120
85-120° fraction	4	30
Hexane	1.7	80
	0.25	20
Toluene	20	110
	1.9	20

The solvents listed in Table 5-12 are representative of the type of solvents which would be most applicable for use in a process for removal of elemental sulfur from coal. All contain only elements (sulfur and/or carbon) naturally present in the coal matrix so that retention by the coal of a small amount of solvent would not result in a contamination problem. All are available in large quantities on a commercial basis and are relatively inexpensive. With the exception of carbon disulfide, the solvents listed in Table 5-12 have a large solubility differential over the temperature range from approximately room temperature to near boiling. Thus, each of these solvents could extract sulfur near the boiling point, then crystallize sulfur on cooling. The solvents are incapable of dissolving more than a minor amount of the coal matrix, at the short extraction times and relatively low temperatures required, with the exception of the coal tar oil, which is an excellent solvent for coal.

Current experimentation has concentrated on the use of toluene for the extraction of sulfur from coal treated by the Meyers Process (13,16) since this solvent has a relatively large differential sulfur solubility as a function of temperature, is a poor coal solvent and also forms an attractive azeotrope with water which allows the toluene to be utilized for azeotropic drying of coal prior to sulfur extraction (2.7 g of water/g of toluene at 85°C).

Thus, coal can be "dried" with a minimum weight of toluene. The sulfur obtained in cooling toluene is crystalline and for Lower Kittanning coal was approximately 78% pure (13). The small amount of coal which had dissolved in solution during extraction co-precipitated with sulfur.

Of course, a key consideration in the concept of using a solvent for removal of elemental sulfur involves the amount of retention of solvent by the coal. The retention of toluene by desulfurized coal was evaluated (13,16). Toluene wet coal was dried in the laboratory under careful material balanced conditions. It was found that an upper limit of 2% w/w toluene (and most probably much less) was retained on the coal as a result of drying in laboratory apparatus.

3. CHEMICAL REACTION

The recovery of elemental sulfur from coal by chemical reaction requires a) conversion of sulfur to a volatile, soluble or otherwise coal separable form, b) reagent selectivity to sulfur, and c) a simple regeneration scheme.

Sulfur and hydrogen combine directly at 150-200°C (28) to form hydrogen sulfide. Since this temperature is well below that which will cause reaction of hydrogen with the coal matrix, hydrogen either with or without a solvent medium should be effective in removing elemental sulfur from coal. Ideally, water-wet coal could be treated with hydrogen in an autoclave to convert sulfur to H_2S. The H_2S would then be recovered and converted to elemental sulfur. The coal product could be used directly as power plant fuel after any necessary drying.

A number of chemicals are available which react with elemental sulfur to form soluble sulfur compounds. These include ammonia and ammonium hydroxide which form mixtures of polysulfides, sulfides, thiosulfates and sulfites. However, a mixture of products is undesirable as an unnecessary complication of reagent regeneration.

The most attractive chemical reaction scheme for removal of elemental sulfur from coal would be one in which a single product is formed (to minimize regeneration and recycle problems), and from which elemental sulfur could be recovered easily. Haver and Baker (29) report that over 99% of the elemental sulfur remaining in an oxidized copper ore residue may be extracted in five minutes residence time using a 21% aqueous ammonium sulfide solution. This solution dissolves up to 250 g/l of sulfur at room temperature. The resulting polysulfide solution may be decomposed by heating at 80-100°C to give ammonia and hydrogen sulfide gases and coarse rhombic sulfur crystals (over 99.5% pure). The ammonia and hydrogen sulfide gases recombine in water to regenerate ammonium sulfide solution.

Although this technique has not yet been investigated, it could be applied for the recovery of elemental sulfur from coal. Wet coal, which has been leached with ferric sulfate solution, could be treated with aqueous ammonium sulfide solution at or near room temperature causing conversion of the adsorbed elemental sulfur to ammonium polysulfide (Eq. 16). Residual iron sulfate salt will also dissolve into the solution. After separation of the filtrate, the solution would be heated to 80-100°C, decomposing the ammonium polysulfide (Eq. 17) to elemental sulfur solid and distilling off ammonia and hydrogen sulfide vapors. The elemental sulfur would then be filtered from the solution and the ammonia and hydrogen sulfide recombined with water to regenerate ammonium sulfide (Eq. 18).

$$(NH_4)_2S + S_X \rightarrow (NH_4)_2S_{X+1} \tag{16}$$

$$(NH_4)_2S_{X+1} \rightarrow S_X + 2NH_3 + H_2S \tag{17}$$

$$2NH_3 + H_2S \rightarrow (NH_4)_2S \tag{18}$$

Dissolved iron sulfate could pose a problem as it might react with ammonia to form iron ammonium sulfates or metathesize to precipitate iron sulfide. Thus, it would probably be advantageous to thoroughly wash the coal prior to treatment. The coal filter cake, which would be free of elemental sulfur, could be rinsed with water to remove residual ammonium sulfide, or alternatively, the coal could be given a cursory wash then heated to a temperature of about 40-80°C to decompose any residual ammonium sulfide to ammonia and hydrogen sulfide (plus a small amount of sulfur which could be collected and recycled).

Thus, the coal would not require drying as a part of the sulfur vaporization or the sulfur solvent extraction methods.

B. Rejection of Product Iron and Sulfate

The leaching of pyrite in coal with aqueous ferric sulfate results in the dissolution of additional iron and sulfate into the leach solution. For the observed molar sulfate/sulfur ratio of 1.5 it is necessary to remove 1.0 mole of iron and 1.2 moles of sulfate from the leach solution for each mole of pyrite reacted in order to maintain constant composition of the leach solution.

Ferric ion may be regenerated by oxygen or air according to Eqs. 19 or 20.

$$2.4\ O_2 + 9.6\ FeSO_4 + 4.8\ H_2SO_4 \rightarrow 4.8\ Fe_2(SO_4)_3 + 4.8\ H_2O \tag{19}$$

$$2.3\ O_2 + 9.2\ FeSO_4 + 4.6\ H_2SO_4 \rightarrow 4.6\ Fe_2(SO_4)_3 + 4.6\ H_2O \tag{20}$$

The form of the process products varies with the degree of regeneration performed. Thus, Eqs. 19 and 20 lead to the overall process chemistry indicated by Eqs. 21 and 22, where the products are mixtures of iron sulfates and elemental sulfur (Eq. 21) or ferrous sulfate, sulfuric acid and elemental sulfur (Eq. 22).

$$FeS_2 + 2.4 \, O_2 \rightarrow 0.6 \, FeSO_4 + 0.2 \, Fe_2(SO_4)_3 + 0.8S \qquad (21)$$

$$FeS_2 + 2.3 \, O_2 + 0.2 \, H_2O \rightarrow FeSO_4 + 0.2 \, H_2SO_4 + 0.8S \qquad (22)$$

Several options exist in product recovery. Iron sulfates may be recovered as pure solids by stepwise evaporation of a spent reagent slipstream with ferrous sulfate being recovered first because of its lower solubility. Ferrous sulfate may be recovered by crystallization and ferric sulfate or sulfuric acid removed by liming spent reagent or spent wash water slipstreams. Alternatively, the spent leach solution may be fully reduced to ferrous sulfate by reduction with scrap iron and ferrous sulfate precipitated as above.

Orsini (30) studied the precipitation of ferrous sulfate from spent iron sulfate leach solutions of varying acidity (Table 5-13). Leach solution, concentrated to 40% of its original volume, precipitated 20 to 60% of the ferrous sulfate content at $100^\circ C$. Higher acidity favored ferrous sulfate precipitation, while ferrous sulfate is more soluble at lower temperatures. Thus, leach solution can be separated from reacted coal at temperatures less than $80^\circ C$, heated to precipitate ferrous sulfate and increase the Y value, then recycled to the leacher.

TABLE 5-13

Ferrous Sulfate Solubility in Spent Iron Sulfate*
Leach Solutions (30)

Treatment	Initial H_2SO_4 conc, % w/w	Precipitation T, $^\circ C$	Solution Y	Solution Fe^{+2} conc, % w/w
None	2 or 4	–	0.6	2
Concentrated**	2	80	0.6	5
Concentrated**	2	100	0.7	4
Concentrated**	4	60	0.6	5
Concentrated**	4	80	0.7	4
Concentrated**	4	100	0.8	2

*Initial Fe_T concentration of 5% w/w.

**Spent leach solution was concentrated to 40% of original volume.

Watanabe (31) reported that iron sulfates, specifically ferrous sulfate, can be precipitated from spent pickle liquor by addition of 35 to 38% w/w acetone at a temperature of approximately $80^\circ C$. The product ferrous sulfate contains only about 1% w/w adsorbed acetone. Bartholomew (32) found that waste pickle liquor containing sulphuric acid and ferrous sulfate can be concentrated by evaporization of water (utilizing a submerged combustor) to precipitate a readily filterable ferrous sulfate · 1.5 H_2O.

Kramarsic *et al* reported that ferrous sulfate can be precipitated from waste pickle liquors by super cooling to -12 to -16°C giving crystalline hydrated ferrous sulfate as a solid product (33). Iron and sulfate can be directly removed from the leach solution by neutralization with lime to produce iron oxide and calcium sulfate.

Thus, there are a number of methods for the removal of iron and sulfate from the various leach liquor process streams in appropriate ratios of ferrous, ferric and sulfate according to the particular process design contemplated, degree of regeneration and specific SO_4^{-2}/S ratio. Iron sulfates may be stored as such for sale or may easily be converted to highly insoluble basic iron sulfates (by air oxidation) or calcium sulfate (by low-temperature solid phase reaction) for disposal.

C. Fate of Minor and Trace Elements

Up to 50% or more of the weight of coal produced by U.S. mining operations can consist of mineral matter. More than 95% of the mineral matter associated with normal marketable coal consists of species belonging to the shale, kaolin, sulfide and chloride groups (*cf* Chapter 2). When coal is burned, these mineral constituents form an ash residue composed chiefly of compounds of silicon, aluminum, iron and calcium with smaller quantities of almost every element found in the earth's crust.

Some of these minerals are soluble in acidic or basic aqueous solutions or may be brought into solution upon oxidation. Thus, many of the possible chemical extraction solutions discussed in this book would be expected to dissolve a part of the mineral (ash) portion of the coal matrix.

The aqueous ferric sulfate solutions used in the Meyers Process are both acidic and oxidizing and it would be expected that some portion of the coal-ash component would dissolve in the leach solution. Thus, excess ash removal, over that calculated for pyrite dissolution, was studied. As part of a survey of the applicability of the Meyers Process for desulfurization of U.S. coal (34), thirty-four run-of-mine U.S. coals from the Appalachian and interior basins through the western area were treated with ferric sulfate under standard conditions.

The coals studied had initial ash contents of 7.55 to 49.28% and extracted ash contents of 3.37 to 43.46%. In each case there was a small to significant excess ash removal over that which could be accounted for by removal of pyrite (Table 5-14). Coal from individual mines showed as little as 0.2-1.0% (w/w of coal) excess ash removal, while 2-3% excess ash removal was typical for the Appalachian and Interior Basin coals and Western coals averaged 3.9%. There appeared to be no correlation by seams but it was shown that excess ash removal increased with increasing ash content. This was most apparent with Appalachian coals where high ash run-of-mine coals experienced more than three times the excess ash removal of low ash run-of-mine coals. Typically, coal cleaning will reduce the ash content by at least 50% (relative), thus the excess removal by the Meyers Process should be reduced by roughly this fraction when cleaned coal is extracted.

TABLE 5-14

Average Excess Ash Removal (% w/w)

Region		Seam		Ash Content*	
Appalachian (23)** 2.4***		Sewickley**	2.6	Low (7)**	1.2***
		Pittsburgh (7)	2.7		
		Freeport (3)	2.4	Medium (8)	2.1
		Kittanning (5)	2.3		
				High (8)	3.8
		Others (5)	2.4		
Eastern & Western Interior (7)	2.6	Herrin No. 6 (2)	1.9	Low (4)	2.4
		Illinois No. 5 (4)	2.8	Medium (3)	2.9
		Des Moines No. 1 (1)	3.3	High (0)	–
Western (4)	3.9			Low (3)	3.7
				Medium (1)	4.5
				H•h (0)	–

*Low Ash, 0-15%; Medium Ash, 15-25%; High Ash, >25%.

**Number of mines in sample.

***Average excess ash loss in % w/w on coal.

Most of the excess ash removal can probably be accounted for by dissolution of basic inorganic compounds by the relatively acidic ferric sulfate solution (pH 1-2). These include basic salts of sodium, potassium, calcium, magnesium, and iron. These compounds tend to be rejected from the process stream along with iron sulfates during the crystallization process (Section B).

In addition, a number of trace elements are removed by the ferric sulfate leaching technique (34). Trace elements are removed either by a process of neutralization or, as in the case of the various acid insoluble sulfide mineral constituents of coal (arsenopyrite, chalcopyrite, sphalerite, etc.), by an oxidative reaction similar to that operable for the dissolution of iron pyrite.

Hamersma *et al* (34) studied the removal of trace elements during pyrite leaching at 100°C for 20 U.S. coals by analysis of treated and untreated coal. Analyses were performed for antimony, arsenic, beryllium, boron, cadmium, chromium, copper, fluorine, lead, lithium, magnesium, mercury, nickel, selenium, silver, tin, vanadium, and zinc. He also measured trace element removal which could be obtained by physical cleaning by analyzing float-sink treated coal samples.

Hamersma found that both chemical leaching and physical cleaning processes have the ability to remove trace elements. The results are shown in Figure 5-16b in cumulative fashion. Six elements, B, Be, Hg, Sb, Se, and Sn yielded negative or inconclusive results due to low levels or analysis difficulties and were not plotted. Although the results varied greatly from coal to coal in respect to the elements extracted and the degree of extraction, some general conclusions were advanced by Hamersma:

(1)   As, Cd, Mn, Ni, Pb, and Zn are removed to a significantly greater extent by the Meyers Process

(2)   F and Li are partitioned to a greater extent by physical separation procedures

(3)   Ag and Cu are removed with a slight preference for float-sink separation, and

(4)   Cr and V are removed for both processes with equal success.

A note of caution should be made regarding the relationship of float-sink data to actual physical cleaning results for trace element removal. The float-sink procedure involves contacting coal with a chlorinated organic solvent, which picks up water from the coal and may hydrolyze on repeated use to form small amounts of aqueous hydrochloric acid. The acid could then dissolve some trace elements leading to artificially high values for trace element removal. Trace element removal values are shown for some specific mines from each region in Table 5-15.

The removal of trace elements during simultaneous leach and regeneration (see Section II.D.) was studied with special regard to both minor and trace elements (16). Analysis of the fresh and recycled leach solution (Table 5-16) showed that minor elements (which, as their salts, can comprise 10-20% of the inorganic coal structure) such as calcium, sodium and magnesium are also removed from coal by ferric sulfate solution (in addition to many trace elements). Chloride ion present in the coal also dissolves into the leach solution.

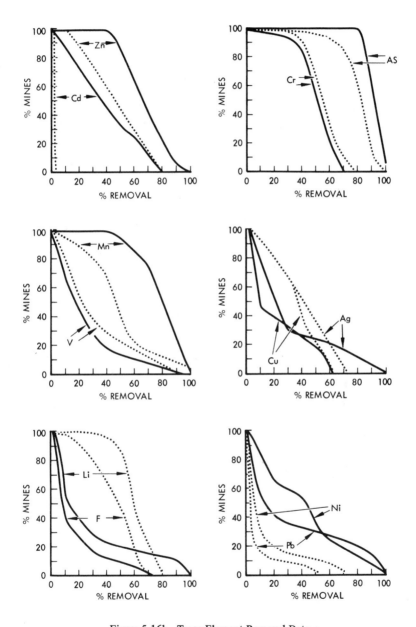

Figure 5-16b. Trace Element Removal Data

TABLE 5-15

Trace Element Removal – Coal Analysis (23)

	Removal (% w/w)		
Element	Appalachian*	Eastern Interior**	Western***
As	81 ± 7	90 ± 17	–
Cr	60 ± 14	71 ± 5	–
Mn	87 ± 7	77 ± 7	93 ± 3
Ni	–	65 ± 10	58 ± 5
Pb	99 ± 2	98 ± 1	93 ± 11
Zn	50 ± 21	84 ± 1	–

*Jane mine, Lower Freeport seam.
**Eagle No. 2 mine, Illinois No. 5 seam.
***Colstrip mine, Rosebud seam.

TABLE 5-16

Trace Element Removal – Solution Analysis (16)

	Solution Concentration (ppm)	
Element	Starting**	Recycle
As	<1	19
Be	0.06	0.9
Ca	5	318
Cd	0.07	0.14
Cl	37	175
Cu	0.09	64
Cr	6	30
Mg	6	238
Mn	28	30
Na	4	36
Ni	6	154
Pb	0.44	0.65
Zn	0.78	52

*Lower Kittanning seam coal.
**Reagent grade ferric sulfate dissolved in distilled water.

It was possible to obtain a precipitate on cooling of the recycle solution which analyzed better than 99% calcium sulfate with trace amounts of magnesium, silicon and strontium. It would thus be expected that calcium sulfate and the sulfate salts of other minor and trace elements could either be precipitated separately or co-precipitate from the recycle solution along with iron sulfate salts during leach solution recycle (see Section B). Chloride ion would also tend to co-precipitate as mixed surface salts.

It is likely that, as the elemental composition of the leach solution builds up to a steady state in a continuous recycle situation, the solubility coefficient of many of the elements would be such that no further removal would take place. For example, when the solubility of calcium sulfate in iron sulfate solution (e.g., 0.1619 g/100 cc water at 100°C) is exceeded, no further calcium will dissolve into the solution. Therefore, the leach process itself may not remove large amounts of calcium from coal feed at steady state. Thus, lime or limestone in coal should result in neutralization of the leach solution and some precipitation of calcium sulfate and iron oxide although it may be possible to confine this precipitation to leach solution crystallization step. The initial sulfur removal capability of the process is reduced to the extent that precipitation on coal occurs. However, subsequent water or acid washing would remove this problem. It is expected that the solubility coefficient of other elements would probably not be exceeded during leaching as they are present in small amounts and would tend to co-precipitate with iron sulfate to produce essentially technical grade salts.

IV. APPLICABILITY TO U.S. COALS

The applicability of a process for the removal of pyritic sulfur from coal for air pollution control is not only a function of the degree of sulfur removal attained and the amount of residual sulfur in treated coal, *i.e.*, pollution control standards, but also is strongly dependent on the selectivity of the desulfurization reaction (*cf* Section II), the rate of pyritic sulfur removal and the ability of the process to be integrated with existing coal cleaning technology, *i.e.*, process economics.

As discussed in Chapter 1, the distribution of sulfur content between pyritic and organic sulfur forms in coal is highly variable as a function of a) the various coal basins, b) seams within these basins, and c) petrographic and geological location within a particular seam. Other properties of coal, such as ash content and composition and chemical and physical properties, also vary according to these factors. Therefore, in order to evaluate the applicability of a coal treatment process, it is necessary to obtain information regarding the variation of processing features as a function of the variations in coal composition.

The investigation of Hamersma *et al* on the application of the Meyers Process to a group of U.S. coals involved the selection of 35 coal mines based on the following criteria: a) selection of representatives of the widest possible variety of coal beds, coal regions, and coal rank, and b) high production and reserves. Coal mines from New Mexico and Montana through Iowa, Illinois, Ohio, Alabama, Tennessee, Kentucky, Pennsylvania and West Virginia were tested, representing a wide range of U.S. production and reserves, and including all major coal basins and many of the largest mines in the United States. Emphasis was placed on coal mines in the Appalachian basin, since over 60% of the current U.S. production comes from this region and there are more than 300 x 10^9 tons of reserves, yet only 10-15% of the coal now mined can meet the Federal Standards for New Stationary Sources (34). The sulfur forms distribution of these coals is presented in Figure 1-1 (Chapter 1). The results of these processing studies are presented in the following sections on sulfur removal, selectivity and heat content changes, and process combination with coal cleaning.

A. Sulfur Removal

The Meyers Process is effective for sulfur removal over a wide range of processing conditions, e.g., 100-130°C, coal top sizes of 1/4 in. to 200 mesh, pressures from ambient to 100 psig, and both with and without simultaneous leach and regeneration. A set of reaction conditions amenable to simple laboratory testing within the above range of variables was selected as a technique for surveying the applicability of the process to U.S. coals. The testing was conducted at approximately 100°C, atmospheric pressure, with the leach solution periodically changed to maintain a high Y value (*cf* Section II.B.) and at top sizes of 100 or 200 mesh.

A summary of the sulfur removal results obtained on run-of-mine (uncleaned) coal sampled from 32 U.S. coal mines is shown in Table 5-17. Many of these coals contained 1, 2 and up to 5% w/w pyritic sulfur. Coal was sampled from three additional mines, but these contained 0.2% w/w pyritic sulfur or less and so were not further evaluated. The sulfur content of each coal before and after extraction is shown in columns 4 and 5 (from left). Pyritic sulfur removal was generally in the 90-99% range (column 7). Because a large portion of U.S. coal production is physically cleaned prior to shipment for removal of noncombustible rock and varying amounts of pyrite and carbon, float-sink fractionation was performed to define the relative utility for each coal of physical cleaning and chemical desulfurization. The data in Table 5-17 shows that: a) the Meyers Process removes 83-99% of the pyritic sulfur content of the 32 coals studied, resulting in total sulfur content reductions of 25-80%, b) twelve (38% of the coals) are reduced in sulfur content to the 0.6-0.8% sulfur levels generally consistent with Federal Standards for New Stationary Sources (38) as well as many states standards, c) in all cases the Meyers Process removes significant to very large increments of sulfur over that separable by physical cleaning (*cf* Chapter 1). In two cases, the Mathies and North River mines, coal cleaning results in a sulfur content increase in the float product.

TABLE 5-17

Summary of Pyrite Removal Results*

Mine	Seam	State	% Sulfur w/w in Coal**			Meyers Process pyrite conversion (% w/w)
			Initial	Fe$_2$(SO$_4$)$_3$ extracted	After efficient*** float sink	
Navajo	Nos. 6,7, and 8	N. Mex	0.8	0.6	–	89
Kopperston	Campbell Creek	W. Va	0.9	0.6	0.8	92
Harris Nos. 1&7	Eagle No. 2 Gas	W. Va	1.0	0.8	0.9	94
Colstrip	Rosebud	Mont	1.0	0.6	–	83
Warwick	Sewickley	Pa	1.4	0.6	1.0	92
Marion	Upper Freeport	Pa	1.4	0.7	1.2	96
Mathies	Pittsburgh	Pa	1.5	0.9	1.7	95
Isabella	Pittsburgh	Pa	1.6	0.7	1.5	96
Orient No. 6	Herrin No. 6	Ill	1.7	0.9	1.4	96
Lucas	Middle Kittanning	Pa	1.8	0.6	0.7	94
Jane	Lower Freeport	Pa	1.8	0.7	0.8	91
Martinka	Lower Kittanning	W. Va	2.0	0.6	0.8	92
North River	Corona	Ala	2.1	0.9	2.2	91
Humphrey No. 7	Pittsburgh	W. Va	2.6	1.5	1.9	91
No. 1	Mason	E. Ky	3.1	1.6	2.3	90
Bird No. 3	Lower Kittanning	Pa	3.1	0.8	1.5	96
Williams	Pittsburgh	W. Va	3.5	1.7	2.3	96
Shoemaker	Pittsburgh	W. Va	3.5	1.7	3.6	96
Meigs	Clarion 4A	Ohio	3.7	1.9	2.8	93
Fox	Lower Kittanning	Pa	3.8	1.6	2.0	93
Dean	Dean	Tenn	4.1	2.1	3.0	94
Powhattan No. 4	Pittsburgh No. 8	Ohio	4.1	1.9	3.3	99
Eagle No. 2	Illinois No. 5	Ill	4.3	2.0	2.9	94
Star	No. 9	W. Ky	4.3	2.5	3.0	91
Robinson	Pittsburgh	W. Va	4.4	2.2	3.0	97
Homestead	No. 11	W. Ky	4.5	1.7	3.2	93
Camp Nos. 1&2	No. 9	W. Ky	4.5	2.0	2.9	99
Ken	No. 9	W. Ky	4.8	2.8	3.5	91
Delmont	Upper Freeport	Pa	4.9	0.8	2.1	96
Muskingum	Meigs Creek	Ohio	6.1	3.2	4.4	94
Weldon	Des Moines No. 1	Iowa	6.4	2.2	3.9	92
Egypt Valley	Pittsburgh No. 8	Ohio	6.6	2.7	4.6	93

*Coal pulverized to -100 to -200 mesh; extracted for 6-23 hrs with 0.5M Fe$_2$(SO$_4$)$_3$ solution at 102°C, washed then extracted with toluene and dried.

**Dry, moisture-free basis.

***1.90 float material, 14 mesh x 0.

B. Selectivity and Heat Content Changes

The selectivity of the ferric sulfate reaction with the pyrite content of coal was evaluated independently by measuring excess ferric ion consumption over that required for pyrite removal and any decreases in coal heat content over that estimated for pyrite removal. Most coals show a net heat content increase after ferric sulfate leaching and removal of generated elemental sulfur (Table 5-18) because pyrite is a low heat content (2,000 Btu/lb) coal component and ash content removal in excess of that associated with pyrite is obtained. Heat content increases of 1.1-13.2% were observed for Appalachian coals and for 5 of 6 of the Interior Basin coals. However, one of the Western coals showed a heat content decrease because of a significant reaction of the ferric sulfate with the coal matrix.

TABLE 5-18

Summary of Heat Content Changes and Excess Ferric Ion Consumption[*,**]

Coal Mine	Basin	Heat content change				Total Fe[++] expt.[**]
		Dry basis		Dry mineral matter Free basis		
		Btu/lb	% w/w	Btu/lb	% w/w	Excess Fe[+3] (mM/g of coal)
						Total Fe[++] calc.
Warwick		+ 753	+ 8.7	- 197	- 1.3	1.36 / 2.31
Muskingum		+ 564	+ 5.1	- 502	- 3.4	2.42 / 1.45
Egypt Valley No. 21		+ 912	+ 8.6	- 297	- 2.0	0.88 / 1.20
Powhattan		+ 817	+10.2	+ 34	+0.2	1.21 / 1.33
Isabella		+1096	+13.2	+ 2	0	1.05 / 1.79
Mathies		+ 870	+10.7	+ 210	+1.4	1.79 / 3.04
Williams		+ 574	+ 4.4	- 167	- 1.1	1.24 / 1.40
Hymphrey No. 7		+ 318	+ 2.3	- 219	- 1.4	1.00 / 1.43
Robinson Run	Appalachian Basin	+ 802	+ 6.2	- 243	- 1.6	1.20 / 1.27
Shoemaker		+ 661	+ 7.0	- 208	- 1.4	1.94 / 1.71
Delmont		+1096	+10.0	- 260	- 1.6	0.93 / 1.16
Marion		+ 674	+ 6.1	+ 21	+0.1	0.47 / 1.35
Jane		+ 460	+ 3.9	- 242	- 1.5	0.62 / 0.79
Lucas		+ 433	+ 3.2	+ 28	+0.2	1.43 / 1.74
Bird No. 3		+ 949	+ 9.0	- 234	- 1.5	0.92 / 1.21
Fox		+ 201	+ 1.5	- 572	- 3.7	0.72 / 0.89
Martinka		+ 566	+ 7.6	—	—	0.98 / 1.48
Meigs		+ 817	+ 9.0	- 253	- 1.7	2.95 / 1.92
Dean		+ 556	+ 4.6	- 115	- 0.8	1.98 / 1.51
No. 1		+ 287	+ 2.2	- 254	- 1.7	1.69 / 1.90
Kopperston		+ 383	+ 3.5	- 613	(- 3.8)	0.67 / 1.97
Harris Nos. 1&2		+ 142	+ 1.1	- 283	- 1.8	1.12 / 2.53
North River		+ 630	+ 8.2	—	—	0.97 / 1.48
Orient No. 6		- 129	- 1.2	- 916	- 6.2	4.77 / 4.03
Homestead		+ 331	+ 2.8	- 614	- 4.2	5.70 / 2.24
Eagle No. 2	Interior	+ 835	+ 7.9	- 415	- 2.8	1.83 / 1.47
Camp Nos. 1&2		+ 635	+ 5.7	- 314	- 2.2	1.20 / 1.34
Ken		+ 590	+ 4.9	- 405	- 2.8	3.92 / 1.96
Star		+ 342	+ 2.8	- 614	- 4.2	7.85 / 3.09
Weldon		+ 733	+ 6.2	- 864	- 6.0	11.31 / 1.55
Navajo	Western	+ 303	+ 3.0	- 638	- 4.8	6.66 / ***
Colstrip		- 270	- 2.3	- 1072	- 8.2	19.09 / ***

*These values are the averages of replicate 23 hr runs, except where noted.

**The calculated values are based on a sulfate:sulfur ratio of 1.5.

***These values have not been calculated because the low initial pyrite makes them meaningless.

These coal heat content increases are a positive economic benefit to those shipping and using coal. However, a true picture of the process selectivity and Fe^{+3} oxidation of the organic coal matrix is obtained by examining the dry mineral matter free heat content changes (columns 5 and 6). Here it can be seen that heat content losses are observed up to 8% for Western coals and 2-6% for the Eastern and Western basin coals, and heat content changes of +1.4 to -3.7% are measured for the Appalachian basin coals. The mineral-matter-free heat content increase noticed for two of the Appalachian coals is simply a measure of the accuracy of the Btu determinations and is not a chemical reduction phenomenon. The heat content change averaged for all three groups of coals is: Appalachian coals, a loss of $1 \pm 1.2\%$; Interior coals, a loss of $4 \pm 1.5\%$; and Western coals a loss of 6.5%. Although the Interior and Western coals showed ash-free heat content losses because of oxidation by ferric sulfate, these coals can still be considered candidates for desulfurization via the Meyers Process, since approximately 5-8% of the heat content of coal is needed to provide heat for drying and solvent recovery in a process plant (*cf* Section V). Thus, the measured heat content losses translate into an efficient form of internally generated process heat. A statistical analysis of the experimental uncertainties and calculation consumptions for Appalachian coals indicates that heat content loss for the Appalachian region must be considered nil.

A more accurate method for determination of selectivity of the ferric sulfate reaction with pyrite in a coal matrix is obtained by consideration of excess ferric ion consumption (columns 7 and 8 of Table 5-18). When these results are averaged by region (Table 5-19), it can be seen that the excess ferric ion reaction decreases as the rank of the coal increases. Western coals have low rank and an open pore structure which provides an abundance of active sites for oxidation. Eastern Interior basin coals have a higher rank but still have a reasonably open pore structure that allows substantial reaction, while Appalachian basin coals, though similar in rank to the Eastern Interior, have a relatively closed pore structure and show very little reaction with ferric ion. These results follow the degree of metamorphism of the coals. Comparison of the average excess ferric ion consumption against the measured heat content losses (Table 5-19) shows the consistent relationship between excess ferric ion consumption and heat content loss.

Thus, Appalachian coals undergo a highly selective reaction with ferric sulfate in which essentially only pyrite is oxidized. An additional indication of the non-reactivity of the Appalachian coal organic matrix with ferric sulfate solution is obtained by considering free swelling index (fsi) data (Table 5-20). This test, which relates to the coal caking properties, shows that there is little or no change of fsi for Appalachian coal after treatment by the Meyers Process. Contrastingly, mild weathering of coal (an air oxidation process) causes a significant decrease in fsi.

TABLE 5-19

Average Heat Content Losses and Ferric Ion Consumption

Coal Basin	Heat content loss (Btu/% heat content loss)	Average excess Fe^{+3} consumption (mM/g)	Excess Fe^{+3} consumption (mM/g/% heat content loss)
Appalachian	172	1.3 ± 0.6	1.3
Interior	148	5.2 ± 3.5	1.3
Western	128	10.8 ± 6.3	1.5
Average	149 ± 22		1.4 ± 0.1

TABLE 5-20

Free Swelling Index of Appalachian Coals
Treated by the Meyers Process

Coal Mine	Fsi	
	Initial	Extracted
Egypt Valley	4	4
Humphrey No. 7	8	8-1/2
Jane	6-1/2	5
Fox	6	7
Martinka	1-1/2	0
Dean	5-1/2	3-1/2
Kopperston	7	5-1/2
Harris Nos. 1 & 2	7	7

C. Combination with Coal Cleaning

As previously noted, the Meyers Process is particularly applicable for desulfurization of Appalachian coal. The total sulfur and pyritic sulfur content data for desulfurization of the raw coal output of the 22 Appalachian coal mines surveyed is shown in Figure 5-18. It can be seen that the total sulfur content of many of these desulfurized run-of-mine coals is within the range which would meet the Federal New Source Performance Standards (4) of 0.6-0.9% w/w sulfur, provided their heat content was in the range of 10,000-15,000 Btu/lb. However, some of these coals were uncleaned, contained considerable ash, and had a relatively low heat content. Some typical analyses of run-of-mine Appalachian coal feed products are shown in Table 5-21. Although the heat content increases for the coals shown range from 3-9% because of pyrite and ash removal, only Mines 5 and 6 meet the Federal Standards for New Stationary Sources of 0.6 lb sulfur/10^6 Btu after chemical desulfurization.

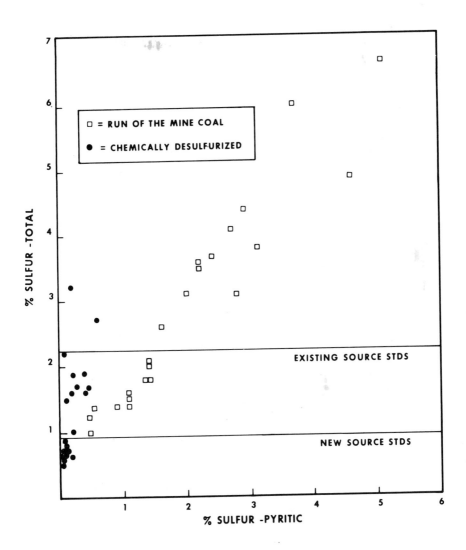

Figure 5-18. Desulfurization of Appalachian Coal

TABLE 5-21

Typical Analyses* of Raw Coal Feed and Chemically Desulfurized Product

Mine/State	% Sulfur		% Ash		Heat content (Btu)	
	Raw	Desulf'd	Raw	Desulf'd	Raw	Desulf'd
Delmont/Pa	4.9	1.0	27.18	20.44	11,012	12,108
Bird No. 3/Pa	3.1	0.8	30.23	24.17	10,551	11,500
North River/Ala	2.1	0.9	49.28	42.84	7,693	8,323
Martinka/W. Va	2.0	0.6	49.25	43.46	7,552	8,138
Lucas/Pa	1.8	0.6	8.68	6.32	13,451	13,884
Marion/Pa	1.4	0.7	26.40	22.61	11,046	11,720

*Dry basis

Physical cleaning has also generally been unable to accomplish sulfur reduction in the range necessary to meet the standards (see Chapter 2). The median sulfur content of the Appalachian coal mines sampled and treated by the Meyers Process is 0.6% lower than the same coals treated for "deep" physical cleaning (14 x 0 mesh, 1.60 float material). Further, while the heat content recovery is 99 ± 1% for the Meyers Process on Appalachian coals, the heat content recovery for physical cleaning as described above has a theoretical maximum of 93 ± 2%, and in practice is significantly less. These heat content losses accumulated by physical cleaning appear as a high ash, low Btu reject when discarded from coal preparation plants which is both a pollution and a safety problem in Appalachia.

The ultimate application of the Meyers Process to meet Federal standards can be greatly improved by performing some type of minimal precleaning of the raw coal prior to desulfurization in order to optimize the economics of sulfur reduction.

It is apparent that most of the remaining coals in Table 5-22 could be reduced by chemical desulfurization to meet the Standards for New Stationary Sources provided that a precleaning step is utilized. For example, float-sink testing data simulating simple prewashing (3/8 in. x 0, 1.90 float) obtained on the remaining four coals (Table 5-22) was used to calculate the pyritic sulfur reduction needed for chemical desulfurization of precleaned coals to meet the New Source Standards. It can be seen that physical cleaning alone would not reduce the total sulfur content of the four coals to levels required by the standards but because of ash removal the heat content of three of the remaining coals rises. Thus, only 56-86% removal of the remaining pyritic sulfur is necessary in order to meet the standards. This degree of removal is far less than current Meyers Process design criteria of 90-95% removal, which allows a very large reduction in leach residence time for process design.

TABLE 5-22

Clean Coal* – Chemical Desulfurization Needed to Meet Standards

Mine	Total S (% w/w)	Pyritic S (% w/w)	Total S Content to meet std** (% w/w)	Heat Content (Btu)	Pyritic sulfur removal needed to meet std
Delmont	2.82	2.38	0.84	13,900	83%
Bird	1.28	0.71	0.88	14,700	56%
North River	2.18	1.27	0.75	12,500	–
Martinka	1.31	0.92	0.73	12,200	63%

*3/8 x 0 in. float at 1.90 sp gr.

**Calculated: $S_{std} = \dfrac{0.6 \times \text{ht content}}{10^4}$.

V. ENGINEERING DESIGN AND COST ESTIMATIONS

Plant design and cost estimation studies for the chemical removal of pyritic sulfur from coal by the Meyers Process have been performed by a number of organizations. Some of the process variations which have been engineered and tested include the following:

1) Leaching and regeneration temperatures from 70 to 130°C

2) Leaching and regeneration in the same vessel and in separate vessels

3) Removal of generated elemental sulfur by vaporization or solvent extraction

4) Air vs oxygen for regeneration

5) Feed coal top sizes of 1/4 in., 14 mesh and 100 mesh

6) Slurry solids content from 20-33%

7) Oxygen partial pressures of 1-10 atmospheres.

All of the above conditions have been found to be effective and their utilization involves economic tradeoffs. Three major design studies have been performed to date by TRW Systems and Energy (16), Dow Chemical, U.S.A. (36), and the Exxon Research and Engineering Co. (37).

The TRW design and cost studies emphasize two basic processing approaches: the processing of coarse shippable coal (e.g., 1/4 inch top size), and the processing of moderately fine coal (14 mesh top size). Major emphasis was placed on fine coal processing. TRW's engineering study is the most recent of the studies, and incorporates the latest process improvements such as a) improved reaction rates, b) use of moderately fine coal rather than very fine coal (100 mesh top size), and c) removal of generated elemental sulfur by vaporization rather than solvent extraction.

The Dow and Exxon engineering studies make use of earlier data, which is summarized in Section I.D.2., and which is essentially limited to the processing of very fine coal (100 mesh top size) and removal of generated elemental sulfur by solvent extraction.

Additional, and necessarily less complete engineering studies were performed earlier in the process development and are not described in this discussion (38,39,40,41).

The results of the TRW, Dow and Exxon engineering studies are presented in the following three sections, entitled TRW Engineering Studies, Dow-Midland Engineering Studies and Exxon Engineering Studies. The engineering approaches developed in these studies should have partial applicability for other chemical coal desulfurization technologies, especially those in Chapter 6.

A. TRW Engineering Studies

A summary of the engineering studies performed at TRW by L.J. Van Nice and M.J. Santy are presented in the following three sections entitled 1) Suspendable Coal Processing Design and Cost Studies, 2) Coarse Coal Processing Design and Cost Studies, and 3) Projected Process Economics. These results, which were published (16) by Van Nice, are liberally quoted, edited, and paraphrased in the following discussion.

1. SUSPENDABLE COAL PROCESSING DESIGN AND COST STUDIES

Van Nice and Santy defined suspendable coal as coal of a small enough particle size that it may be processed as a substantially uniform slurry with moderate mixing energy. Although no rigid top size specification can be given, it appears that coals with top sizes up to approximately 8 mesh may be classed as suspendable. Bench-scale experiments were therefore conducted using 14 mesh and 100 mesh top size coals as representative of suspendable coal (Section II.D.3.). (Both of these sizes are often referred to as fine coal in order to differentiate them from the coarse coal which will be described in Section 2.)

The conceptual full scale process design for the chemical removal of pyritic sulfur from fine coal devised by Van Nice is described in Sections a and b below. The first gives the design basis, which relies heavily on the bench-scale experimental data, but also incorporates information provided by equipment vendors and data obtained from the literature. The second section presents the baseline design with capital and operating cost estimates. A number of major tradeoffs were examined by Van Nice and Santy in arriving at the baseline design. The reader is directed to the literature reference (16) for a complete discussion of these tradeoffs.

a. Design Basis for Suspendable Coal

Processing coal to remove pyritic sulfur using aqueous iron sulfate involves four major process sections, each containing several unit operations. The four process sections are: the reactor section, the washing section, the sulfur removal section, and the sulfate removal section.

The reactor section, which includes mixing and solution regeneration, has three main process requirements:

1) Providing mixing and wetting of ground coal with the aqueous ferric sulfate leach solution and raising the slurry to the operating temperature and pressure.

2) Providing the residence time and reaction conditions which remove a nominal 95% of the pyrite originally contained in feed coal.

3) Providing the residence time and reaction conditions which regenerate the ferric sulfate solution from the spent ferrous sulfate leach solution.

The washing section, which includes several stages of coal washing and coal dewatering, has two main process requirements:

1) Providing for contact of the leach solution-wet coal with a minimum quantity of wash water to remove water soluble iron sulfates.

2) Providing for separation of coal from the leach solution and the wash water.

The sulfur removal section, which removes both sulfur and excess water from the product coal, has three main process requirements:

1) Providing conditions such as heat or solvent contact to remove elemental sulfur from the processed coal.

2) Providing the thermal environment necessary to reduce the moisture level of the coal to the desired value.

3) Providing the means for recovery of the by-product elemental sulfur for subsequent marketing, storage or disposal.

The sulfate removal section, which removes excess iron sulfate from the recycle leach solution, has four main process requirements:

1) Providing for the removal of iron sulfate from the aqueous spent leach solution by crystallization and/or neutralization.

2) Providing for the recovery of wash water from the wash section effluents.

3) Providing for maintaining the correct acid level by neutralizing excess acid if required.

4) Providing for separation of the by-product iron sulfate and neutralization product from the recycle streams.

Specific information and data for the steps or operations which are important to the process design are presented in the following paragraphs. These data rely heavily on the information given in Section II.D.3., but also include additional qualitative observations made during the bench-scale experimental efforts.

Mixing — The bench-scale effort (16) demonstrated that there is a critical aspect to the mixing operation beyond simply surface wetting of the particles and suspension in the leach solution. Preparing the slurry can be readily accomplished with mixing times of 15 minutes or less, but severe foaming of the slurry will occur when pressurized and heated.

The mixing time for a high rank, high ash, dry coal should be between 30 and 60 minutes at the normal boiling point of the solution if subsequent foaming is to be avoided. Lesser times may be possible with moist or low rank coal. The quantity of foam produced tends to decrease with increasing coal particle size and with lower solids content in the slurry. These are secondary parameters, however, and are not of major importance in the process design.

Leach Reaction — The net overall reaction between pyrite and the ferric sulfate leach solution is represented by Eq. 22.

$$FeS_2 + 4.6\ Fe_2(SO_4)_3 + 4.8\ H_2O \longrightarrow 10.2\ FeSO_4 + 4.8\ H_2SO_4 + 0.8S \quad (22)$$

$\Delta H = -55\ Kcal/g\text{-mole}\ FeS_2 = -0.10\ MM\ btu/lb\text{-mole}\ FeS_2$ reacted

The reaction rate was found to have a second order dependence on both the fraction of pyrite (or pyritic sulfur) in the coal and the fraction of the total iron in the leach solution which is in the ferric ion form. The leach rate at temperatures of interest is represented by Eq. 23.

$$r_L = \frac{-d\ [Wp]}{dt} = K_L\ [Wp]^2\ [Y]^2 \quad (23)$$

where

$[Wp]$ = wt% pyrite in dry coal at time t

$[Y]$ = fraction of iron as ferric ion at time t

K_L = leach rate constant (a function of temperature and Wp).

K_L is independent of total iron concentration at least in the immediate vicinity of 3-5% total iron. Physical considerations such as increased solubility density and viscosity and the limited solubility of ferrous sulfate in the ferric sulfate solution become increasingly important to the design of the pyrite leacher when the total iron concentration approaches 10%.

The leach rate constant as a function of temperature can be adequately represented by Eq. 24.

$$K_L = A \times \exp(-E/RT) \tag{24}$$

where:

E = 11,100 cal/mole

R = 1.987 cal/mole - $^{\circ}K$

T = temperature in $^{\circ}K$

A = a function of size, temperature and Wp.

For 14 mesh top size coal at atmospheric pressure and temperatures between $70^{\circ}C$ and the solution boiling point (about $102^{\circ}C$), the value of A is 2.9×10^5 for all values of Wp. At temperatures between $110^{\circ}C$ and $130^{\circ}C$ under oxygen and steam pressure up to about 150 psig, the value of A is 7.4×10^5 when Wp is large (above 1.7) and 2.9×10^5 when Wp is small (below 1.2). Additional refinement of the data in the transition region would be desirable, but an adequate representation of the data can be obtained by a linear decrease in A from 7.4×10^5 for Wp = 1.7 to the A value of 2.9×10^5 for Wp = 1.2.

For 100 mesh top size coal, the ranges of applicability for A appear to be the same, but its value is about 20% higher. Thus at low temperature or low Wp, the value of A is 3.5×10^5 while at higher temperature and high Wp the value is 8.9×10^5.

These leach rate constants have been defined only for the high pyrite, high ash Lower Kittanning coal used in the bench-scale programs. They should not be applied, without experimental verification, to other coals or coal seams. The Lower Kittanning coals investigated in the bench-scale programs had starting values of Wp between 6 and 8. The transition to a lower rate constant thus occurs at about 75-85% pyrite removal. Based on a single test of a coal with a lower ash and low starting Wp (Marion Mine, upper Freeport seam, Wp = 1.7), a high leach rate was found at least to 80 or 90% removal (i.e., Wp about 0.2).

Regeneration — The leach reaction produces both ferrous sulfate and sulfuric acid which must be processed for continuous recycle operation. For each mole of pyrite reacted, 9.6 moles of ferrous sulfate may be regenerated to maintain the acid at a constant level. This gives by-products for disposal of 0.2 moles of $Fe_2 (SO_4)_3$, 0.6 moles of $FeSO_4$ and 0.8 moles of elemental sulfur. Alternately, regeneration of 9.2 moles of ferrous sulfate can be considered if some acid is neutralized to give by-products of 1.0 mole of $FeSO_4$, 0.2 moles of H_2SO_4 and 0.8 moles of elemental sulfur (see Section III.B.). The choice of the extent of regeneration should be made on the basis of the by-product removal steps included in the process design.

The regeneration reaction is shown in Eq. 25.

$$1.0 \ FeSO_4 + 0.5 \ H_2SO_4 + 0.25 \ O_2 \rightarrow 0.5 \ Fe_2(SO_4)_3 + 0.5 \ H_2O \qquad (25)$$

$$\Delta H = -18.6 \ Kcal/g\text{-}mole \ FeSO_4 = -0.0335 \ MM \ btu/lb\text{-}mole \ FeSO_4$$

If hydrolysis of a portion of the ferric sulfate to iron oxide should occur as shown in Eq. 26,

$$Fe_2 (SO_4)_3 + 3 \ H_2O \rightarrow Fe_2O_3 + H_2SO_4 \qquad (26)$$

then additional acid neutralization or regeneration of ferrous ion would be required to remove the acidity produced from the hydrolysis reaction. The extent of hydrolysis at temperatures below 250°F appears to be small, but at higher temperatures there is some evidence of precipitation of ferric oxide and possibly a low hydrate or anhydrous ferrous sulfate. The hydrolysis products and/or precipitates formed at 265°F were found to redissolve slowly in ambient temperature spent leach solution and do not remain as permanent products. No data was obtained above 265°F.

The regeneration rate was found to be second order in the molar concentration of ferrous ion over the range of ferrous concentration from 100% to less than 1% of the total iron. The rate is shown in Eq. 27.

$$r_R = \frac{-d \ [Fe^{+2}]}{dt} = K_R [Fe^{+2}]^2 [O_2] \qquad (27)$$

where:

$[Fe^{+2}]$ = concentration of ferrous ion, mole/liter

$$[O_2] \quad = \quad \text{oxygen partial pressure, atm}$$

$$K_R \quad = \quad 1.836 \text{ liters/mole-atm-hour at } 248^\circ F$$

Over the range of temperatures studied ($212^\circ F$ to $265^\circ F$), the rate constant was found to vary exponentially with temperature (Eq. 28).

$$K_R = 40.2 * 10^6 \exp(-13,200/RT) \tag{28}$$

which gives:

Temp, $^\circ$F	($^\circ$C)	K_R, liters/mole-atm-hour
212	(100)	0.74
230	(110)	1.18
248	(120)	1.84
266	(130)	2.79

The ferric sulfate regeneration rates were obtained under conditions where oxygen in the form of minute air or oxygen bubbles was dispersed throughout the ferrous sulfate solution. Thus, all of the solution was continually saturated with oxygen at the partial pressure of oxygen present in the regeneration gas. At bench-scale, the minute bubbles were formed by pumping a portion of the liquid in turbulent flow ($N_{Re} > 3000$) through a pipe whose length was 50 or more times its diameter. Gas containing oxygen was added to the liquid in an amount ranging from less than 1% to greater than 10% by volume at flow conditions. The method is very similar to aeration equipment used to reduce the biological or chemical oxygen demand of chemical plant effluent streams, except that ferric sulfate regeneration is conducted at higher temperatures.

Separation — The major separation step requires treated, fine coal to be separated from the spent leach solution. The four principal methods which could be employed are hydrocyclones, centrifuges, filters and thickeners. Suspendable coal has a large fraction of particles smaller than 100 microns in diameter and, in general, hydrocyclones are not useful for particle sizes below several hundred microns. Centrifuges would require very high power input and recycle rates to separate the coal from the leach solution because of the fine particle size and the small liquid-solid density difference. Filtration is applicable, but for slurries less than 30 or 35% solids, the filter area requirements increase rapidly. Typically, a 10% solids slurry needs more than ten times the filter area needed for a 35-55% slurry. Thickeners have been used on commercial scale to remove coal fines from water and other aqueous media. Data for similar density solutions and coal sizes were reported in Reference 8. It was estimated that a thickener area of about 20 ft^2 per ton/day of coal with an edge depth of about 8 feet would provide an underflow with greater than 35% solids and an overflow containing only a few tenths percent (or less) solids when the feed contains 10-20% of 100 mesh coal. Since the thickener slurry can be maintained near the leach solution boiling point, the time spent in the thickener could be used to carry the leaching reaction to a greater degree of completion and to redissolve any solids formed during regeneration.

Filtration — The two important design values relating to filtration are the filtration rate and the coal "moisture" content. These values are not independent and are both highly dependent on the specific coal and its properties. Generalized correlations reported in the literature were reviewed and a data point was obtained from a filter manufacturer for a bench-scale slurry of the high ash, Lower Kittanning coal. The vendor report (16) showed that rates equivalent to about 25 lb of dry coal/hr/ft^2 were obtained with a 20% slurry of -100 mesh coal in a 5% iron leach solution, and projected a 60% increase in rate for a slurry with a solids concentration of about 33%.

One reported correlation (42) plots rate against a parameter which is the product of the percent ash in the -200 mesh fraction times the square root of the weight percent of the -200 mesh fraction. For the Lower Kittanning sample tested in the above filtration test, the parameter has a value of about 200 (Eq. 29).

$$(25\% \text{ ash}) \times (67\% \text{ of } -200)^{1/2} = 205 \tag{29}$$

Figure 5-19 shows this measured point and the literature data with extrapolations to a typical 14 mesh top size coal. The expected filtration rate for a cleaned 14 mesh top size coal is expected to be near 200 lb/hr/ft^2 in a 33% slurry and about 150 lb/hr/ft^2 in a 20% slurry. For a cleaned 100 mesh top size coal the filtration rate is expected to be about 100 lb/hr/ft^2 for a 33% slurry and about 70 lb/hr/ft^2 for a 20% slurry.

Data on the moisture content of the filter cake was taken by the filter manufacturer at the same time as rate was measured. The projected cake moisture was 40-45%. This is higher than either reported in the literature (26-34%) or found in a typical bench-scale filtration (about 32%) for the high ash, 100 mesh top size coal (16). The literature values are for water wet cakes rather than leach solution wet cakes. The data for both water and leach solution (5-5.5% iron) are summarized in Table 5-23.

In order to provide for an adequate amount of coal washing to remove the sulfate leach solution, it was decided that 50 parts of liquid per 100 parts of dry coal would be used for both leach solution and water on both coal sizes.

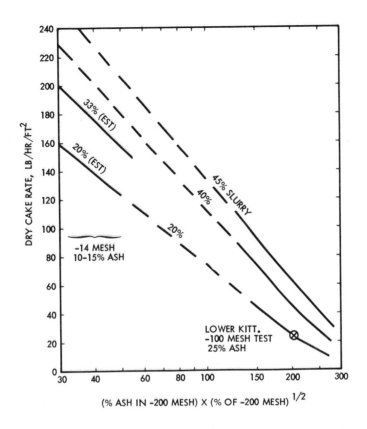

Figure 5-19. Filtration Rate Correlation (16)

TABLE 5-23

Moisture Content of Filter Cakes (16)

Test	Parts of liquid per 100 parts of dry coal	
	100 mesh high ash	14 mesh high ash
Bench scale		
leach solution	45-50	45-50
water	35-40	35-40
Vendor test		
leach solution	65-80	not tested
Reference		
water	35-50	15-25

123

b. Process Baseline Design

A block diagram of the Meyers Process as applied to coal of about 8 mesh top size or finer is shown in Figure 5-20. This block diagram shows the main operations and the interconnections between each of the four process sections. Before discussing the process flow diagrams and mass balance in detail, the block diagram will be described to give a brief process overview.

Reactor Section — Ground coal, with a nominal top size of 14 mesh, is mixed with hot recycled iron sulfate leach solution. After wetting is complete at the solution boiling temperature, the slurry is introduced into a vessel where the majority of the pyrite reaction is accomplished at elevated temperature and pressure. Oxygen is simultaneously added to regenerate the leach solution. Heat of reaction is removed and is used to reheat the recycle leach solution. The slurry is passed to a secondary reactor operated at atmospheric pressure and near the solution boiling temperature where the remaining pyrite reaction occurs.

Wash Section — The iron sulfate leach solution is removed from the powdered coal in a series of countercurrent flow contractors and separators. The slurry from the secondary reactor is first filtered and the cake is washed on the filter. Both the filtrate and wash liquids are sent to the Sulfate Removal Section. The first filter cake is reslurried, filtered a second time and then reslurried with recovered clean water and is finally dewatered in a centrifuge.

Sulfur Removal Section — Moist coal from the centrifuge is flash-dried by high temperature steam which simultaneously vaporizes the elemental sulfur produced in the leach reaction. The dry coal is separated from the hot steam and sulfur vapor stream in a cyclone and cooled to give the clean product coal. The hot sulfur vapor-steam effluent from the cyclone is scrubbed with large quantities of recycled hot water and the liquid sulfur is drawn off to by-product storage. A small part of the hot water is used in the Wash Section, with the remainder circulated to the evaporator.

Sulfate Removal Section — The major function of this section is the evaporation of wash water to concentrate the leach solution for recycle. The filtrate from the wash section and a portion of the spent wash water from the first filter is fed to a triple-effect evaporator which recovers most of the wash water.

The by-product iron sulfate crystals which form in the final stage of evaporation are separated from the concentrated leach solution and stored. The remaining wash water from the first filter is partially neutralized with lime to yield a gypsum by-product. The separated and partially neutralized wash water is combined with the concentrated leach solution from the centrifuge and recycled to the Reactor Section. Overall, the pyrite is reacted with oxygen and water to give ferrous sulfate, sulfuric acid and sulfur. These by-products are removed as shown. The fuel requirement is equal to a few percent of the product coal and makeup water is needed to replace water of crystallization and water vapor loss through the vacuum filters and the vacuum evaporator.

b. Conceptual Design for Commercial Scale

As a result of the process engineering studies and tradeoffs, a baseline flow diagram for a commercial scale plant was prepared by Van Nice. This flow diagram, presented in four sections, is given in Figure 5-21. The corresponding mass balance and stream properties are given in the original report (16). The baseline plant size was chosen equal to 100 tons of dry coal feed per hour equivalent to about 250 MW power plant feed. This size is about the maximum size for a single train based on available commercial equipment. The following sections discuss the primary plant components.

Feed and Mixer - Crushed coal, nominally 14 mesh top size, is fed from feed hopper A-1. The coal is assumed to have 3.2% pyritic sulfur and 10% moisture on a dry basis; thus, the total solids feed rate is 110 tons per hour (TPH) at room temperature, assumed to be 77°F. The coal feed, Stream 1, is brought to the mix tank, T-1, by conveyor, C-1, and introduced through the rotary feed valve, RV-1. Recycled leach solution, Stream 4, at its boiling point (215°F) is introduced to the first mixer stage after first passing through the gas scrubber SP-1. Steam, Streams 2 and 3, is needed to raise the feed coal from 77°F to the 215°F mixer temperature. Approximately 5.6 TPH of atmospheric pressure steam is required to heat the coal while 6.5 TPH is available from the flash drum, T-2. It is possible that the steam would actually be added to the enclosed conveyor to provide heated coal with an effective 15.6% moisture content. The excess 0.9 TPH would be vented through SP-1 along with any flash steam formed in Stream 4.

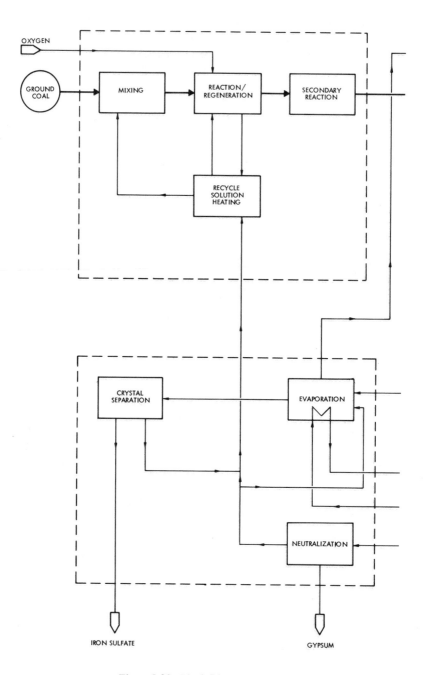

Figure 5-20. Block Diagram for Fine Coal

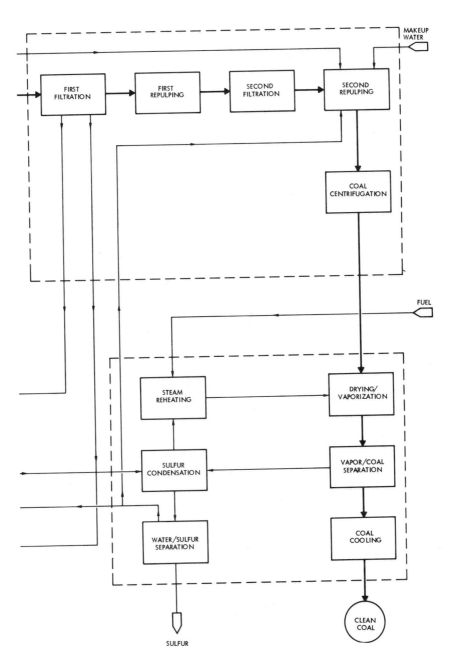

Figure 5-20. Block Diagram for Fine Coal (Continued)

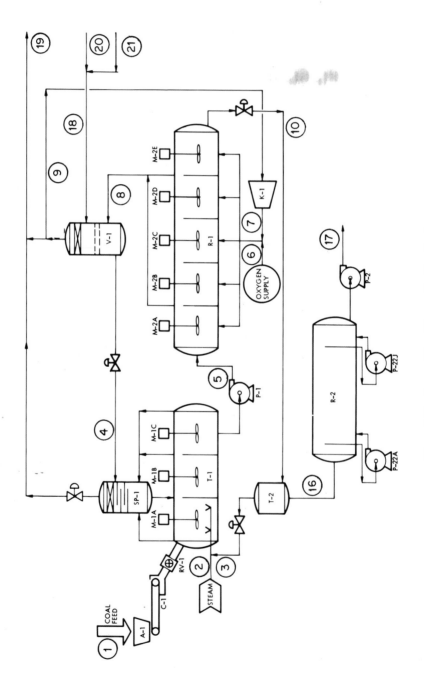

Figure 5-21. Process Flow Diagram for Fine Coal

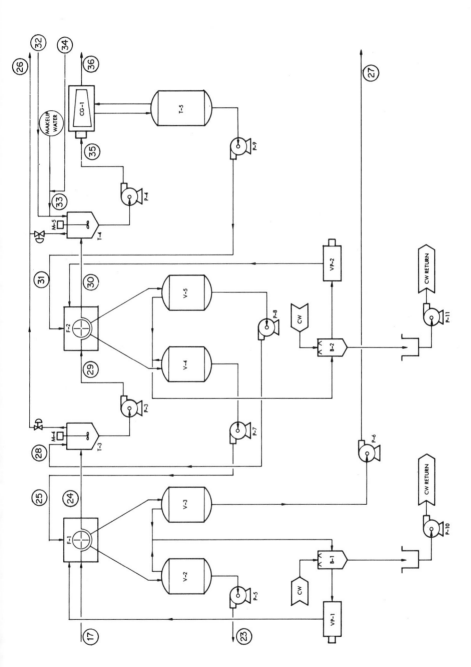

Figure 5-21. Process Flow Diagram for Fine Coal (Continued)

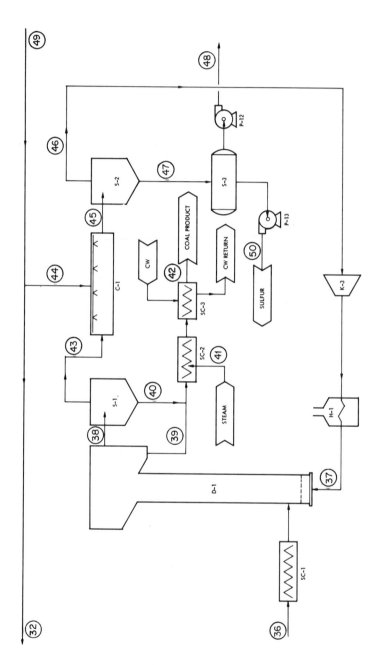

Figure 5-21. Process Flow Diagram for Fine Coal (Continued)

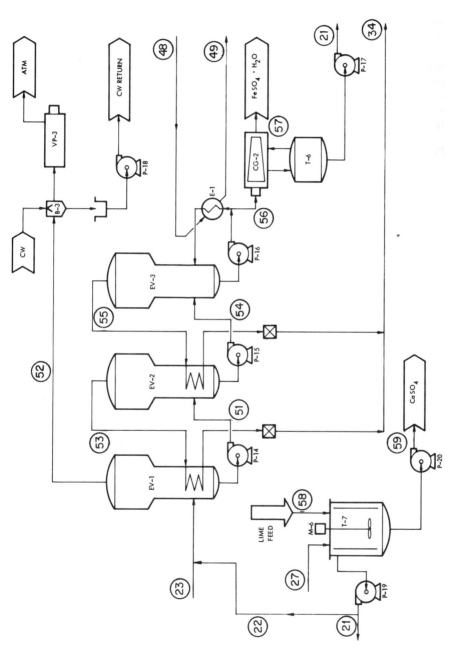

Figure 5-21. Process Flow Diagram for Fine Coal (Continued)

The mixer vessel T-1 was sized for three stages of mixing at 0.25 hours per stage. Under the design constraint that the vessel is 75% full, the cost model used for vessel sizing found a field fabricated vessel 18.7 feet in diameter by 32.9 feet long has minimum cost. The selected vessel size (18 x 36) gives three stages each about 12 feet long and 12.6 feet deep with slightly less than 15,000 gallons in each stage. Any foam generated during coal wetting is broken down and the entrapped air is scrubbed in SP-1 by the returning leach solution. The actual air flow through SP-1 is very low and will probably not exceed the air in the bulk coal (50 cubic feet per minute).

Primary Reactor — The fully wetted and de-aerated coal slurry from the mixer is pumped by slurry pump P-1 (Stream 5) into the first stage of the primary reactor R-1. Both removal of pyrite and oxidation of ferrous to ferric iron sulfate occur in this reactor. A five-stage reactor was selected since the cost model showed the minimum cost field-fabricated vessel had length-to-diameter ratios near five. Under the design constraint that the reactor have five stages and operate about 85% full, the cost model found a reactor 25.9 feet in diameter by 127.7 feet long operated at 15 psi of oxygen was minimum cost. The selected vessel size (26 x 125) gives five stages each about 25 feet long by 23 feet deep and holding about 80,000 gallons of slurry. At the residence time of 1.5 hours per stage, a temperature of $250^{O}F$ and an oxygen partial pressure of 15 psi, the pyrite is reduced to 88% of the original level and the leach solution is regenerated to a Y (ferric iron to total iron ratio) of 0.83 in the primary reactor.

Oxygen Loop — Excess oxygen saturated with steam and containing an equilibrium level of inert gas (mainly argon) leaves the primary reactor in Stream 8. The gas is contacted with returning leach solution, Stream 18, in a knock-out drum, vessel V-1. The leach solution is warmed to $215^{O}F$ (Stream 4) by condensing steam from the oxygen stream. The gaseous effluent, which was assumed to have a $50^{O}F$ approach to the feed leach solution, is split to give a small vent Stream 19 and a recycle oxygen Stream 9. The vent rate is selected to maintain the inert gas at the design level, namely 5% on a dry basis. The recycle oxygen is compressed by K-1 to the reactor feed pressure. Makeup oxygen, Stream 6, is added to balance the oxygen used for regeneration in R-1 and that vented to remove inerts.

Assuming 15 psi oxygen pressure, the gas pressures in reactor R-1 at $250^{O}F$ are as follows:

Oxygen	15.0 psia
Inert Gas	0.8 psia
Steam	27.7 psia
	43.5 psia (28.8 psig)

Since the recycle gas must also overcome the liquid head in the reactor (about 13 psi), the control valve/injector drop (about 10 psi) and other line losses, the recycle compressor was sized to provide a 25 psi pressure increase. For the baseline case this results in a 300 horsepower compressor operating at a 1.58 compression ratio and a reactor feed pressure of 53.8 psig.

Flash Steam — The heat of reaction and regeneration is accommodated in three ways: temperature of the liquid is increased between the mixer and reactor R-1; heat is lost from the insulated walls of the mixer and reactors; and water is evaporated from the solutions. Part of the steam (13.4 TPH) is removed from the recycle oxygen to provide an isothermal primary reactor R-1 at 250°F, and part of the steam (6.5 TPH) is removed by flash drum T-2 in dropping the slurry from reactor R-1 (250°F) to reactor R-2 (215°F). The heat is almost entirely utilized in heating the feed coal and the recycle leach solution.

Secondary Reactor — The secondary reactor, R-2, is operated near the atmospheric boiling point with a residence time of 36 hours. During this time, additional pyrite is removed from the coal to provide an overall pyrite removal of 95%, while the Y of the solution is decreased to a value of 0.68 in the reactor effluent. The low value of Y is desired to provide sufficient ferrous sulfate for removal as the by-product iron form. The cost model found the minimum cost reactor was 27.9 feet in diameter by 465.9 feet long. Since the length is excessive, the reactor is made up of three vessels each 28 feet in diameter and 160 feet long. The reactors contain no internal stages, but have circulating pumps to avoid large vertical concentration gradients from occurring in the solution. The slurry from the secondary reactor, Stream 17, is pumped by P-2 to the first filter F-1.

Coal Washing — Bench-scale experience with removal of the sulfate leach solution from coal shows that the solution may be treated as two types. Surface solution is readily removed by flushing with water or may be readily displaced by a more dilute wash solution. Part of the solution in the pores of the coal particles requires a definite residence time to reach equilibrium with the bulk or surface liquid. Thus, the coal washing section consists of filtration, washing on the filter, equilibration with dilute solution, a second filtration and wash, equilibration with wash water and dewatering in a centrifuge.

First Filter — Coal slurry from the secondary reactor, Stream 17, containing approximately 33% solids is fed to a 12-foot diameter by 24-foot long rotary vacuum filter, F-1. The filtrate from vacuum receiver V-2 is pumped, P-5, to the sulfate removal section, Stream 23. Dilute wash solution from the second filter, Stream 25, is used to wash the filter cake and displace the surface solution on the coal particles. This sulfate rich wash solution is pumped, P-6, from the vacuum receiver V-3 to the sulfate removal section, Stream 27. Vacuum is provided by a 3000 standard cubic feet per minute (SCFM) vacuum pump,

VP-1, which is vented back to the enclosed filter F-1. The vapors and gases removed from the vacuum receivers, V-2 and V-3, passed through a barometric condenser B-1 before entering the vacuum pump. In B-1 most of the flash steam is condensed and enters the cooling water loop where it is pumped to the cooling water tower by P-10.

First Stage Repulping — The washed filter cake from the first filter, Stream 24, and dilute wash water from the second filter are fed to a stirred tank, T-3. This 40,000 gallon tank is operated about three-fourths full to give an average residence time of 30 minutes in order to equilibrate pore solution with the bulk liquid. The slurry, Stream 29, is pumped, P-3 to the second stage filter. Any gases introduced with the cake are vented to the scrubbing system, Stream 26.

Second Filter — The partially washed slurry, Stream 29, containing approximately 33% solids is filtered and washed on a second filter of the same size and type as the first filter. Filtrate is pumped, P-7, from the vacuum receiver V-4 to the first filter wash, Stream 25. Wash water for the second filter, Stream 31, is obtained from the centrate receiver. The partially spent wash water is pumped, P-8, from the vacuum receiver V-5 to the first stage contactor. The vacuum is provided by vacuum pump, VP-2, operating through the barometric condenser B-2.

Second Stage Repulping — The washed filter cake from the second filter, Stream 30, is contacted with water in a 40,000 stirred tank T-4. The wash water is obtained from the dryer, Stream 32; the evaporators, Stream 34; and makeup, Stream 33.

Dewatering — The slurry from the second contactor, Stream 35, is pumped, P-4, to the dewatering centrifuge CG-1. The slurry with approximately 33% solids is separated in the 36-inch diameter by 90-inch long solid bowl centrifuge to provide a dewatered coal, Stream 36. According to vendor literature and discussions, the dewatered coal is expected to have about 15% moisture. The centrate from receiver T-5 is pumped, P-9, to provide the wash for the second filter, Stream 31.

Drying — Coal from the centrifuge, Stream 36, is fed to a flash dryer D-1 by a screw feeder, SC-1. In the dryer the coal is heated to about 450°F by superheated steam, Stream 37, and carried upward to the enlarged top area of the dryer. The larger particles are removed from the dryer, Stream 39, while the fine particles and gas, Stream 38, are fed to a cyclone, S-1. During the drying in D-1 sulfur is also vaporized from the coal (see Section III.A.1.) and is present along with water vapor in the cyclone effluent gas, Stream 43. The fine coal from the cyclone, Stream 40, and coarse coal, Stream 39, are let down to atmospheric pressure by screw conveyor, SC-2, which is back purged with a small quantity of steam to prevent the sulfur containing gas in the cyclone from leaving the system with the coal. The coal is then transported and cooled to product storage temperature, Stream 42, by the screw conveyor SC-3, which rejects heat either to cooling water or to the atmosphere.

Sulfur Removal — The cyclone effluent gas at about 450°F, Stream 43, is cooled by a large spray of water, Stream 44, in gas cooler, C-1. The water is obtained from the return Stream 49 from the sulfate removal section. The gas and liquid Stream 45 cooled to 250°F is separated in cyclone S-2 to give a vapor Stream 46 and a liquid Stream 47. The liquid Stream 47 contains the water fed to the gas cooler, Stream 44; the water vaporized from the coal in the dryer; and the sulfur vaporized from the coal. The liquid is phase separated in vessel S-3. The liquid sulfur by-product, Stream 50, is pumped, P-13, to storage while the hot water, Stream 48, is pumped, P-12, to the sulfate removal section.

Steam Circulation — Saturated steam at 250°F from the cyclone, Stream 46, is compressed by K-3 and reheated by H-1 then fed to the dryer, Stream 37. Compression is accomplished by two 3500 HP series compressors which make up the 10 psi pressure drop around the gas circulation loop. The heater provides nearly 100 million Btu per hour to the steam to supply the heat required to heat the dryer feed, Stream 36, to 450°F and vaporize the water and sulfur. Slightly more than 80 MM Btu/hr are rejected to the hot water loop, Streams 48 and 49, for use in the sulfate removal section while about 15 MM Btu/hr are lost from the equipment and lines or rejected as sensible heat in the hot coal and liquid sulfur. The circulating water is kept in balance by returning a portion of the water, Stream 32, to the wash section equal to the water vaporized from the feed coal, Stream 36.

Neutralization — Sulfate rich wash solution from the wash section, Stream 27, is fed to a stirred tank, T-7, and a lime slurry, Stream 58, is added to neutralize part of the sulfuric acid. The tank is sized for about 10 minutes of residence time and has a baffled settling zone. Gypsum slurry, Stream 59, is withdrawn for disposal and the partially neutralized liquid is removed by pump P-19. A portion of the liquid, Stream 21, is returned to the reactor section while the remainder, Stream 22, is combined with the filtrate, Stream 23, as feed to the triple effect evaporators.

Evaporation — Evaporator EV-1 is operated at partial vacuum (about 0.1 atmosphere) and uses condensing steam from the second evaporator, Stream 53, to evaporate water, Stream 52, in the first evaporator. The evaporated steam is condensed in the barometric condensor B-3 and any residual gas is removed by vacuum pump VP-3. The partially concentrated leach solution, Stream 51, is pumped, P-14, to the second evaporator, EV-2. The second evaporator operates at about 155°F and 0.2 atmosphere using steam from the third evaporator, Stream 55, to evaporate the water, Stream 53. The two condensate streams from the reboilers of the first and second evaporators (Streams 53 and 55) are combined, Stream 34, to provide clean wash water for the wash section. The leach solution from the second evaporator which has been concentrated to 8.3% iron, Stream 54, is at a temperature where the solubility of ferrous sulfate is a maximum and is a solids free solution. This stream is fed to the third evaporator, EV-3, which is operated at atmospheric pressure and at the normal boiling point of the solution. Heat to vaporize water is provided to the reboiler E-1, by the hot water loop from the wash section (Streams 48 and 49). The overhead steam, Stream 55, is used in the second evaporator as previously described. The leach solution in EV-3 is concentrated to a total iron concentration of nearly 12% which exceeds the solubility of ferrous sulfate. Thus, crystalline ferrous sulfate forms in EV-3 and a portion of the slurry, Stream 56, is fed to a centrifuge CG-2 to separate the crystals, Stream 57, from the concentrated leach solution, Stream 20. The concentrated leach solution is pumped, P-17, to the reactor section.

Solubilities — Since the solubility of ferrous sulfate in the presence of ferric sulfate, sulfuric acid and trace ions is not yet completely defined (see Section III.A.1.), the baseline process flows may require some adjustment when pilot scale data have been evaluated. Nevertheless, the planned mode of operation which takes advantage of the reported solubility characteristics of ferrous sulfate in aqueous solution should be applicable. Below about $150^{\circ}F$, the equilibrium crystalline phase is $FeSO_4 \cdot 7H_2O$ which has an increasing solubility with temperature. It reaches a maximum solubility of nearly 60 grams of $FeSO_4$ (anhydrous basis) per 100 grams of water. Above about $150^{\circ}F$ the equilibrium solid phase is $FeSO_4 \cdot H_2O$ which has a decreasing solubility in water with increasing temperature. Both the first and second stages of evaporation are below the saturation limits and are expected to remain solids free. Only the final stage operates as a crystallizer and produces crystalline ferrous sulfate both from a decreased solubility at the higher temperature and from an increased concentration because of evaporation.

c. Process Cost Estimate

Throughout the bench scale development, process costs were frequently reviewed with an objective of focusing experimental effort in the areas of greatest cost sensitivity. The capital cost of equipment required to perform the pyrite leaching must be carefully controlled to maintain a low processing cost per ton of coal product. As will be seen in the capital estimate presented in the following discussion, the major capital cost continues to be in the reactor section of the process. This section of the unit accounts for approximately 48% of the total installed equipment capital cost. The sulfur removal section of the process accounts for 22%, while the wash section represents 17% and the sulfate removal section 13% of the total equipment capital requirements. It therefore becomes apparent that the reactor section of the process represents the most likely area of future process economic gains as the design data base broadens and other innovative process schemes (relative to the reaction section of the process) are evaluated.

As the process development progressed and additional experimental data were obtained, some complications were identified and some process simplifications were demonstrated. The net result is that at the conclusion of this bench scale effort (16), the process for removing pyritic sulfur from coal remains highly attractive and sufficient data has been obtained to provide confidence in the economic viability of the process.

Baseline Capital Cost Estimate

Table 5-24 summarizes the costs of the equipment which was selected and sized to approach the optimum cost for processing the high (3.2%) pyritic sulfur coal to the 95% removal level, based on the conceptual process design and process flow sheet developed in the preceding section. A complete list of the major equipment for the process is given in Reference 16, and is correlated with the equipment of the flow sheet (Figure 5-13).

TABLE 5-24

TRW Suspendable Coal Process Equipment Cost* by Section

Section	FOB Cost $ x 10^6	Installed Cost $ x 10^6
Reactor	3.26	6.36
Wash	1.16	2.28
Sulfur Removal	1.42	2.91
Sulfate Removal	0.97	1.68
Total Estimated Cost	6.81	13.26

*Mid-1975 $

Operating Cost Estimate — The process operating costs including a capital charge for the battery limits plant were estimated. The bases of these estimates were technical literature and informal supplier quotes. Specific sources of information are presented in Reference 16. The total estimated processing cost in mid-1975 dollars is shown in Table 5-25.

TABLE 5-25

Processing Costs

Capital Related Costs:		Annual Cost, $1000
Depreciation – 10% straight line		1,326
Maintenance, insurance, taxes, interest		1,989
Labor:		
Labor, 8 operating positions		1,200
Utilities:		
Electrical power – 7500 KW (25 mil/Kw-hr)		1,500
Cooling water – 20°F rise; 9500 gpm (5¢/1000 gal)		228
Heating – 97 MM Btu/hr; coal, 5T/hr		–
Process water – 110 gpm (25¢/1000 gal)		13
Materials:		
Oxygen 99.5%, 3.9T/hr ($25/T)		780
Lime – .5T/hr ($28/T)		112
TOTAL COST		7,148
Feed coal 100T/hr, 0.8MMT/yr	$8.94/T	
Coal yield (weight basis)	89%	
Coal yield (Btu basis)	94%	

The added cost of energy may also be considered for the baseline coal. If the baseline coal is similar to the Lower Kittanning coal utilized in the laboratory studies, it will contain about 20% ash and have a heating value of 12,300 Btu/lb as fed. After processing, the coal will be 89% recovered, have 14% ash, and have a heating value of 12,900 Btu/lb. With feed coal prices at $15.00/T, the feed costs 62¢/MM Btu. After processing, the available energy costs $1.04/MM Btu.

Based on the current conceptual process design, it was concluded that a broad spectrum of Eastern coals can be processed at costs of about $9.00 per ton. It was assumed in developing these costs that the pyrite removal plant is coordinated with a power plant which will have the principal offsite facilities such as coal grinding facilities, rail service, change-house, offices, etc.

2. COARSE COAL PROCESSING DESIGN AND COST STUDIES

a. Concept Development and Schematic

The physical characteristics and laboratory-derived leach reaction rates of coarse coal required that Van Nice (16) develop a completely different processing concept than that conceived for fine (suspendable) coal. It was of course understood that one of the primary virtues of coarse coal processing is that the product is shippable, handleable and storable by conventional means. In terms of processing, coarse coal: 1) requires less preleaching preparation; 2) eliminates possible requirements for product pelletizing or briquetting where the processing plant is remote from the utility; and 3) allows the potential application of a wide range of conventional physical separation systems to augment the productivity/ investment ratio of the Meyers process. Since only a small amount of process data was available at the time of Van Nice's study (see Section II.B.), the resultant design was highly conceptual in nature.

Van Nice found that the major physical characteristic of coarse coal which necessitates modification of the fine coal processing concept is the tendency of the 1/4 in. x 0 coal grind to separate from a slurry. The fines (approximately 48 mesh x 0) are nominally slurryable, while the larger mesh fraction (about 1/4 in. x 48 mesh) settles out quite rapidly. Because of the tendency for the coal and leach or wash liquors to separate, it is more efficient to retain the coal in a fixed position and allow the liquors to pass through the reacting mass (as in a packed bed) and to transport the coal mechanically by means of conveyor systems.

While the basic leaching reaction established during fine coal investigations is fully applicable, the range of particle sizes included in the nominal 1/4 in. x 0 grind gave rise to a corresponding range of reaction rates. The fines react in basic accordance with the rate constants derived in the fine coal evaluations, but the coarse particles are considerably slower. Since there is little coarse coal rate constant advantage offered by the use of pressurized reaction vessels (16), the coarse coal conceptual process must allow for relatively long coal retention periods. The leaching residence time may be shortened by preprocessing the coarse ROM coal through conventional physical cleaning (scalping) which offers a means of segregating a small, extremely pyrite-rich and difficult to leach fraction of coal for direct disposal. Further economic advantages are indicated when the scalped coal is deep cleaned into two fractions, one lean in pyrite that would not require depyritization and one rich in pyrite to serve as feed stream to the

Meyers Process. The recombined product would meet the sulfur standards even though only approximately one-half of the ROM coal was subjected to the Meyers Process.

Leaching for any given length of time will, because of the difference in leach rate constants, result in the fines fraction being more completely reacted than the coarse fraction. The conceptual design takes advantage of this characteristic by allowing the fines to slurry in the leach solution circuit to a maximum of 10% by weight. This approach assures the achievement of a very high degree of conversion (>95%) in a portion of the coal feed, thereby advantageously lowering the bulk product sulfur content.

Van Nice and a co-worker, C.M. Murray, found that a total of four reactor schemes are potentially viable. These include an above-ground batch reactor, a lined pit batch reactor, a continuous countercurrent reactor, and a continuous cocurrent reactor. All of these possibilities were conceived as atmospheric pressure units in which only the leach reaction occurs; because of the relatively easy coarse coal-from-liquid separation, the regeneration reaction can be accomplished in a separate vessel optimized for that purpose.

The four reactor configurations were simulated with computer models and evaluated in detail. It quickly became apparent that the above-ground batch reactor scheme would be the most expensive and complex of the group and it was therefore eliminated from consideration on that basis. It was also found that the continuous countercurrent reactor requires a smaller vessel and a lower leach solution throughput than the cocurrent reactor; therefore, the continuous cocurrent reactor scheme was also eliminated.

Both the continuous countercurrent and batch pit reactor schemes were expanded in order to complete conceptual flow diagrams for further study. The process, as described in the following paragraphs, could incorporate either reactor type and could be preceded by a physical separation train. The conceptual design is presented schematically in Figure 5-22. A brief description of the conceptual design follows, and details of the key elements are presented in subsequent subsections.

Coal is fed to the reactor, Stream 1, at a design rate of 100 TPH where it contacts a continuously regenerated stream of leach solution, Stream 2, of Y_{ENTER} = 0.95. Spent leach solution, containing a portion of the fines fraction from the coal feed in addition to the 10% slurry, is withdrawn from the reactor, Stream 3, and blended with the liquid Stream 17 recovered from the coal drain. Stream 17 also contains a portion of the fines fed to the reactor via Stream 1 which is acquired during the drain and rinse cycle. The weak leach solution-coal slurry is bled, Stream 5, at a rate sufficient to maintain the total iron content of 4% and to maintain the fines content of the leach

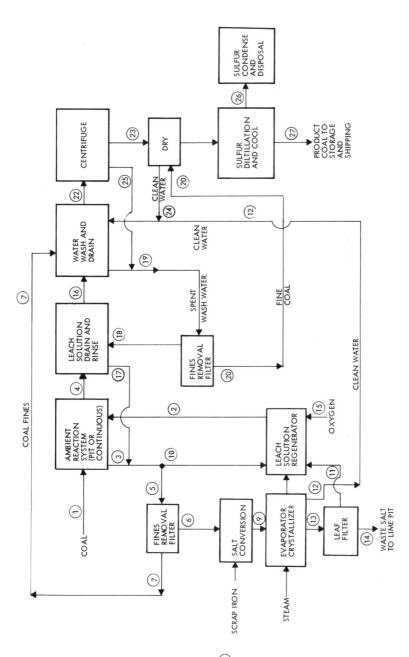

Figure 5-22. Coarse Coal Process Schematic (16)

solution loop at a maximum of 10%. The fines in bleed streams are quantitatively removed by filtration with the filter cake being conveyed, Stream 7, for reinjection into the coal processing sequence at the water wash step. The filtrate is pumped, Stream 6, to a unit which converts the major portion of the Fe^{+3} to Fe^{+2} by iron reduction. The Fe^{+2} rich stream is then sent to an evaporator/crystallizer unit, Stream 9, which produces the clean water used in the coal wash, Stream 12. The ferrous salt is filtered from the crystallizer output, Stream 13, and disposed of through liming, Stream 14. Residual solution, taken as filtrate, is routed to the regenerator, Stream 11, together with the spent leach solution which bypasses the bleed stream diverter, Stream 10. Regeneration of the leach solution is accomplished as outlined in Section 1.a. relative to fine coal processing; oxygen is supplied, Stream 15, to convert the Fe^{+2} ion and sulfuric acid to Fe^{+3} and water. The freshly regenerated leach solution completes the loop through reinjection into the reactor, Stream 2.

The leach solution-wet coal leaving the reactor, Stream 4, is drained on a conveyor and spray-rinsed with fines-free leach solution contaminated water, Stream 18. This rinse water stream is comprised of the overflow from the wash vessel and washed coal drain conveyor; the clean fines carried with Stream 19 are filtered out and the wet cake is conveyed directly to the dryer via Stream 20.

The wash vessel is sized for a nominal one hour residence time to permit transfer of the leach solution trapped in coal pores to the passing water stream, which originates as evaporator and dryer condensates, Streams 12 and 24. Following a conveyor draining step similar to that employed earlier to remove excess leach solution, the coal is transported to a continuous centrifuge, Stream 22, which reduces the total water content to about 20-22% with the centrate going to the fines filter, Stream 25. The coal is conveyed, Stream 23, to a two-stage inert atmosphere dryer/distillation unit which water is quantitatively removed for Stream 24, reused and the elemental sulfur is distilled. The elemental sulfur is cast into blocks for disposal and the coal is cooled for preshipment storage, Stream 27.

A detailed process description is presented in Reference 16.

b. Process Cost Estimate

Van Nice (Reference 16) calculated processing costs, including both estimates of capital requirements and annualized costs, for both the pit and continuous process concepts (*vide-supra*). The dollar basis for calculation was June 1975 in both instances. For purposes of comparison to process costs for fine coal processing (as presented earlier in Section 1.d.), the plant capital requirement was assumed to be that for a "battery limits" facility; that is, the economics presented below do not account for plant off-sites and do not represent a "grass roots" operation.

Coarse Coal Capital Cost Estimate

Process cost estimates are presented for four process configurations. The four approaches — pit, continuous, pit/float-sink combination, and continuous/float-sink combination — were evaluated on the basis of a Meyers Process feed rate of 100 TPH coal containing 3.2% pyritic sulfur.

The objective of front-end physical separation equipment in the float sink combination options is to isolate a low pyrite segment of the total feed which does not require further treatment and to concentrate the pyrite in a "sink" fraction which can be fed to the coarse coal Meyers Process. Float-sink separation data for approximately 400 Appalachian coals were examined. It was found that on the average each of these coals could be separated into two approximately equal weight portions by float/sink separation in a liquid medium with a specific gravity of 1.3. Further, it was noted that in 227 of the coals the float fraction contained less than 1% total sulfur. For these 227 coals the data shows that the 1.3 specific gravity "float" fraction amounts to nearly 55% of the 3/8 inch by zero sample with a standard deviation of about 15%. The average total sulfur content of this float fraction is 0.8% with a standard deviation of 0.1%.

Float-sink equipment trains are presently in widespread use, but use higher specific gravity media to achieve float fractions in the 90% range, thereby minimizing losses and maximizing the salable product. It is apparent that if a separation were made at low specific gravity as suggested above but without the benefit of a Meyers Process for treating the "sink" fraction, the gross effect would be a virtual doubling of coal prices as well as the generation of about five times as much spoil requiring acceptable disposal.

In combining the float-sink treatment with the Meyers coarse coal process it was assumed that a 50-50 float-sink division of the feed would represent a moderately conservative position, based on the analysis described above. Further analysis of these coals indicated that the sink fraction to be used as process feed would contain about 3.8% pyritic sulfur. In a commercial application an economic tradeoff would be made at this point to establish the most advantageous position in terms of the advisability of blending the "float" fraction with the process product (thereby reducing the degree of pyrite removal in the process) versus adjustment of the media specific gravity to control the quantity of "sink" fraction fed to the process. In the conceptual designs considered here, no reblending is considered in order to facilitate direct comparison with the fine coal process.

Table 5-26 presents an equipment list detailing installed equipment costs prepared by Van Nice for 100 TPH pit and continuous coarse coal battery limits process options. The installed equipment capital required for the pit process option is $4.17 million, while that required for the continuous process option is about 50% higher. It should be noted that for the pit and continuous battery limits processes, the reactor sections account for approximately $1.54 million and $3.95 million respectively, or 37% and 60% of the total capital requirement (compared to 48% for the fine coal processing base case). The additional installed capital investment required for the addition of float-sink options is estimated to be $0.42 million.

TABLE 5-26

Coarse Coal Process Equipment Lists

Item	Pit Reactor $1000	Continuous Reactor $1000
Reactor	728	3,175
Regenerator and Exchanger	658	625
Compressor	6	6
Wash Vessel	91	91
Centrifuge	79	79
K.O. Pot	10	10
Regeneration Agitators	106	106
Flash Vessel and Exchanger	21	21
Regenerator Pumps	8	8
Leach Solution Storage	18	24
Leach Surge Tank	8	8
Iron Converison Tank	24	24
Iron Conversion Tank Exchanger	70	70
Iron Conversion Tank Pump	2	2
Fines Filter (Leach Circuit)	65	65
Salt Filter (Plate)	32	32
Evaporator	556	556
Crystallizer	71	71
Fines Cake Conveyors (2)	11	11
Crystallizer Feed/Exit Pumps	4	4
Evaporator Feed/Exit Pumps	11	11
Evaporator/Filter Ejector	12	12
Evaporator Cond. Vessel	4	4
Coal Rinse Conveyor and Sump	62	62
Wash Water Feed/Exit Pumps	6	6
Wash Rack Pump	1	1
Centrifuge Fines Filter	64	64
Wash Vessel Drain Conveyor	30	30
Centrifuge Liquid Exit Pump	1	1
Sulfur Trap	18	18
Evaporator and Dryer Cond. Hold Tank	12	12
Dryer (2 Units)	1,381	1,381
Process Equipment Totals	4,170	6,590

Operating Cost Estimate

The process operating costs have been estimated for three coarse coal options. These are summarized in Table 5-27 and compared to the operating costs for the fine coal process. The operating costs for the coarse coal options range from $2.94 per ton of coal processed to $6.01 per ton. This is comparable to the $8.94 per ton determined for the fine coal processing scheme. However, it must be stressed that the operating costs presented in Table 5-27 are based on process designs with slightly varying bases. Those differences are the following:

1) The pit reactor, continuous reactor and fine coal processes are 100 TPH processing facilities for 3.2% pyritic sulfur coal, while the two float-sink options are 200 TPH units using cleaned coal.

2) The pit/float-sink and continuous/float-sink processes assume the feed coal was precleaned to about 2% pyritic sulfur prior to gravity separation. The resultant gravity separated fractions are assumed to contain 3.8% pyritic sulfur (Meyers Process treated) and less than 0.5% pyritic sulfur (bypassed around the Meyers Process).

The coal yield data (weight basis) presented in Table 5-27 indicates a higher yield for the combined float-sink options (92%) than for the pit and continuous reactor processes options (85%). This is based on the design criteria that all of the material bypassed around the Meyers Process, 50% of the 200 TPH feed, is 100% recovered and blended with coarse coal Meyers Process product. The coal yield in terms of energy recovery is also based on the above stated design criteria.

3. PROJECTION OF PROCESS ECONOMICS

Several Meyers Process options were evaluated by Van Nice (16) to determine their overall process economics, including the cost of off-site equipment. The evaluated processing options included both fine and coarse coal approaches (Sections 1 and 2) integrated into grass roots coal desulfurization facilities. This section presents a description of the various integrated process options, a discussion of the economics evaluation approach, and results of the analysis which in effect yield approximate coal market price requirements for the various options. Table 5-28 summarizes process cost estimates for cases based on utility financing and input coal costs of $20/ton ($0.80/MM Btu). Investor financing and input coal costs of $10/ton and $30/ton were also considered. The bases for all of the results are described in the following sections.

TABLE 5-27

Meyers Process Annualized Costs

| Cost Element – $1000/Yr | Process Option | | | | |
	Pit reactor process*	Continuous reactor process*	Pit/float-sink combination**	Continuous/ float-sink combination**	Fine coal process*
Capital Related Costs					
Depreci- ation – 10%	417	659	459	701	1,326
Maintenance, Insurance, Taxes	626	989	689	1,052	1,989
Labor	900 (6 Positions)	900 (6 Positions)	1,200 (8 Positions)	1,200 (8 Positions)	1,200 (8 Positions)
Utilities					
Electricity (25 mil/kw-hr)	600 (3,000 kw)	600 (3,000 kw)	700 (3,500 kw)	700 (3,500 kw)	1,500 (7,500 kw)
Heating – Coal Equivalent	9.5 (247 MM Btu/hr)	9.5 (247 MM Btu/hr)	9.5 (247 MM Btu/hr)	9.5 (247 MM Btu/hr)	5 TPH (97 MM Btu/hr)
Water- Process and Cooling	458***	458***	458***	458***	241
Materials					
Oxygen 99.5% ($25/Ton)	800	800	800	800	780
Lime ($28/ Ton)	112	112	112	112	112
Scrap Iron ($50/Ton)	280	280	280	280	–
TOTAL	4,193	4,798	4,698	5,303	7,148
Cost/Ton Feed – $/Ton	$5.25/Ton	$6.01/Ton	$2.94/Ton	$3.31/Ton	$8.94/Ton
% Coal Yield (Weight Basis)	85%	85%	92%	92%	89%
% Coal Yield (Btu Basis)	90%	90%	95%	95%	94%

*Requirements for 100 TPH coal feed.

**Requirements for 200 TPH combined coal feed.

***Based on evaporation to reject waste heat.

TABLE 5-28

Meyers Process Estimates (Utility Financing)

Case	Processing cost ¢/MM Btu	Cost of desulfurized fuel-$/MM Btu
Cleaned Fine Coal	50	1.31
ROM Coarse Coal	51	1.32
Physical Separation-Process Pyrite Rich Reject as Fine Coal-Recombine	39	1.20
Physical Separation-Process Pyrite Rich Reject as Coarse Coal-Recombine	33	1.14

a. Processing Option Description

The integrated desulfurization facilities include all battery limits Meyers Process equipment in either fine or coarse coal treatment configuration, plus the required off-sites. The off-sites include such items as:

● Feed and product coal handling and transport equipment

● Physical coal cleaning facilities and size separation equipment

● By-product handling and storage facilities

● Waste treatment (physical cleaning and process generated) and storage facilities

● Process water treatment, storage and pumping facilities

● Cooling water treatment and pumping equipment

● Power and steam generation facilities

● Site office buildings and shop structures

● Other site improvements such as roads, fences, railroad spurs, etc.

It should be noted that the economic evaluations do not include land costs and assume that oxygen is purchased as an over-the-fence utility item (i.e., neither battery limit or off-site equipment include an oxygen plant).

For purposes of economic evaluation, four differing process configuration cases were developed. The central basis for each case was a 100-ton per hour Meyers Process unit. A brief description of each case is presented below and block diagrams containing simplified mass balances for each case are presented in Figure 5-23.

Case 1 – Cleaned Fine Coal Case

Run-of-mine (ROM) coal is physically cleaned and then reduced to 14 mesh top size. The unit feed rate is 120 tons per hour of coal containing about 20% ash and 3 to 4% pyritic sulfur. The cleaning plant refuse (20 tons per hour) is assumed to contain approximately 75% ash; 10 to 14% pyritic sulfur and 11 to 15% coal. The Meyers Process feed consists of 100 tons per hour of coal containing approximately 9% ash and 1.5 to 2% pyritic sulfur. The fine coal Meyers Process configuration utilized in this case is essentially that previously described with the exception that the reaction and filtration associated equipment requirements are significantly reduced due to expected reaction rate improvements (2 to 3 times ROM fine coal processing base case) and lowered ash contents. The product rate is 93 tons per hour of coal containing about 6% ash and 0.1% pyritic sulfur (93-95% removal of pyritic sulfur).

Case 2 – ROM Coarse Coal Case

ROM 1/4 inch top size coal is fed directly to a coarse coal continuous Meyers Process (described in Section 5.2). The coal feed rate is 100 tons per hour and the coal consists of 20% ash and 3 to 4% pyrite sulfur. Processing removes 6 TPH of pyrite and requires 9 TPH of coal to supply process heat. The pyritic sulfur is approximately 95% removed yielding a product rate of 85 tons per hour of coal containing 16% ash and 0.2% pyritic sulfur.

Case 3 – Deep Cleaned Fine Coal with 50% Meyers Process Bypass

ROM coal containing 20% ash and 3 to 4% pyritic sulfur is fed at the rate of 240 tons per hour to a physical cleaning plant and then reduced to 14 mesh top size. The ash discard (40 tons per hour) contains 75% ash, 10 to 14% pyritic sulfur and 11 to 15% coal. The cleaned coal containing 9% ash and 1.5 to 2% pyritic sulfur is fed to a gravity separation unit at a rate of 200 tons per hour. The heavy fraction consists of 100 tons per hour of coal containing 15 to 18% ash and 3 to 4% pyritic sulfur, while the light portion consists of 100 tons per hour of coal with little ash or pyritic sulfur content. The heavy fraction is fed to a fine coal Meyers Process unit which produces 90 tons per hour of product coal containing 10 to 13% ash and 0.2% pyritic sulfur. The Meyers Process product, when blended with the bypassed

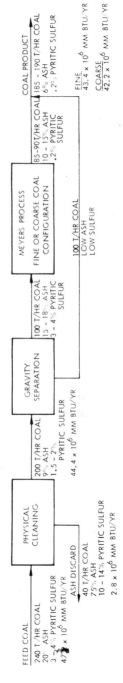

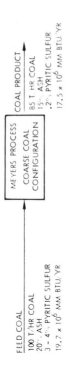

CASE 1 – CLEANED FINE COAL (14 MESH TOP SIZE)

FEED COAL
120 T/HR COAL
20% ASH
3 – 4% PYRITIC
       SULFUR
23.6 × 10⁶
MM BTU/YR

PHYSICAL
CLEANING

100 T/HR COAL
9% ASH
1.5 – 2% PYRITIC SULFUR
22.2 × 10⁶ MM BTU/YR

ASH DISCARD
20 T/HR COAL
75% ASH
10 – 14% PYRITIC SULFUR
1.4 × 10⁶ MM BTU/YR

MEYERS PROCESS
FINE COAL
CONFIGURATION

COAL PRODUCT
93 T/HR COAL
6% ASH
.1% PYRITIC SULFUR
21.3 × 10⁶ MM BTU/YR

CASE 2 – RUN-OF-MINE COARSE COAL (1/4 IN. TOP SIZE)

FEED COAL
100 T/HR COAL
20% ASH
3 – 4% PYRITIC SULFUR
19.7 × 10⁶ MM BTU/YR

MEYERS PROCESS
COARSE COAL
CONFIGURATION

COAL PRODUCT
85 T/HR COAL
15% ASH
.2% PYRITIC SULFUR
17.5 × 10⁶ MM BTU/YR

CASES 3 AND 4, – DEEP CLEANED FINE AND COARSE COAL WITH 50% MEYERS PROCESS BYPASS

FEED COAL
240 T/HR COAL
20% ASH
3 – 4% PYRITIC SULFUR
47 × 10⁶ MM BTU/YR

PHYSICAL
CLEANING

200 T/HR COAL
9% ASH
1.5 – 2% PYRITIC SULFUR
44.4 × 10⁶ MM BTU/YR

ASH DISCARD
40 T/HR COAL
75% ASH
10 – 14% PYRITIC SULFUR
2.8 × 10⁶ MM BTU/YR

GRAVITY
SEPARATION

100 T/HR COAL
15 – 18% ASH
3 – 4% PYRITIC SULFUR

100 T/HR COAL
LOW ASH
LOW SULFUR

MEYERS PROCESS
FINE OR COARSE COAL
CONFIGURATION

85 – 90 T/HR COAL
10 – 15% ASH
.2% PYRITIC SULFUR

COAL PRODUCT
185 – 190 T/HR COAL
6% ASH
.2% PYRITIC SULFUR

FINE
43.4 × 10⁶ MM BTU/YR

COARSE
42.2 × 10⁶ MM BTU/YR

Figure 5-23. Process Economics Case Block Diagram (16)

light coal fraction, yields a combined product stream of 190 tons per hour of coal containing about 6% ash and 0.2% pyritic sulfur (overall 90-95% removal of pyritic sulfur from product coal).

Case 4 — Deep Cleaned Coarse Coal with 50% Process Bypass

Two hundred and forty tons per hour of 1/4 inch top size ROM coal is treated as previously discussed in Case 3. The only difference in the treatment scheme is that a continuous coarse coal Meyers Process configuration is used. For purposes of economic comparisons, the simplified mass balance for Case 4 is assumed to be identical to that described for Case 3. The coarse coal process requires more coal for internal process heat and therefore yields only 185 TPH of product.

b. Process Economics Evaluations

Calculations (Tables 5-29, 30, 31, and 32) were performed at three assumed ROM coal costs and using both utility and investor financing criteria. A detailed discussion of the utility and investor financing economic models is presented in Reference 16. Coal costs of $10 per ton, $20 per ton and $30 per ton were selected since they represent the broad range of currently reported ROM coal costs ($10 per ton at mine mouth and $25 per ton reported delivered price at some TVA plant sites). As indicated by the data, the required market value of the treated coal ranges from a low of 66¢/MM Btu for Case 4 (assuming $10/ton ROM coal cost and utility financing) to a high of $2.39/MM Btu for Case 1 (assuming $30/ton ROM coal cost and investor financing). In all cases, utility financing yielded market costs on the order of 60 to 80% of market costs required when utilizing investor financing.

An equivalent upgrading cost was also determined by Van Nice. The upgrading cost was found by deducting the cost of the dirty energy ($0.41/MM Btu for 20% ash, 3-4% sulfur coal at $10/ton; $0.81 at $20/ton and $1.22 at $30/ton) from the cost shown in Tables 5-29 through 5-32 for the clean energy ($0.66 to $2.39/MM Btu). These upgrading or processing costs, which range from $0.25/MM Btu to $1.17/MM Btu, are presented in Table 5-33. For all cases the processing cost includes ash reduction as well as sulfur reduction. Except for Case 2, physical cleaning was assumed to be coupled with pyrite removal which results in a major reduction in ash from about 20% to about 6%.

TABLE 5-29

Case 1, Cleaned Fine Coal Case

Case product annual energy value, 21.3 x 10⁶ MM Btu/yr	ROM coal cost		
	$10/Ton	$20/Ton	$30/Ton
Capital Related Requirements, $MM			
Battery Limit Capital	8.26	8.26	8.26
Offsite Capital	6.63	6.63	6.63
Overhead and Profit	3.28	3.28	3.28
Engineering and Design	1.49	1.49	1.49
Contingency	2.95	2.95	2.95
Total Plant Investment*	22.61	22.61	22.61
Interest for Construction	3.82	3.82	3.82
Startup Costs	2.89	4.81	6.73
Working Capital (Utility Financing)	2.54	4.57	6.59
Working Capital (Investor Financing)	2.98	5.05	7.12
Total Capital Related Costs (Utility)	31.86	35.81	39.75
Total Capital Related Costs (Investor)	32.30	36.29	40.28
Operating Costs, $MM/Yr			
Raw Material (Coal)	9.60	19.20	28.80
Chemicals (Lime, Scrap)	0.06	0.06	0.06
Supplies	0.50	0.50	0.50
Disposal	0.42	0.42	0.42
Utilities	1.63	1.63	1.63
Labor (13 Positions)	1.62	1.62	1.62
Taxes and Insurance	0.61	0.61	0.61
Total Operating Costs	14.44	24.04	33.64
Required Coal Market Price, $/MM Btu			
Utility Financing	0.84	1.31	1.79
Investor Financing	1.33	1.86	2.39

*Equivalent to a plant capital investment of $79.60/kw.

TABLE 5-30

Case 2, ROM Coarse Coal Case

Case product annual energy value, 17.5×10^6 MM Btu/yr	ROM coal cost		
	$10/Ton	$20/Ton	$30/Ton
Capital Related Requirements, $MM			
Battery Limit Capital	4.20	4.20	4.20
Offsite Capital	4.20	4.20	4.20
Overhead and Profit	1.85	1.85	1.85
Engineering and Design	0.84	0.84	0.84
Contingency	1.66	1.66	1.66
Total Plant Investment*	12.75	12.75	12.75
Interest for Construction	2.15	2.15	2.15
Startup Costs	2.51	4.11	5.71
Working Capital (Utility Financing)	2.06	3.74	5.42
Working Capital (Investor Financing)	2.33	4.05	5.77
Total Capital Related Costs (Utility)	19.47	22.75	26.03
Total Capital Related Costs (Investor)	19.74	23.06	26.38
Operating Costs, $MM/Yr			
Raw Material (Coal)	8.00	16.00	24.00
Chemicals (Lime, Scrap)	0.39	0.39	0.39
Supplies	0.30	0.30	0.30
Disposal	0.64	0.64	0.64
Utilities	1.86	1.86	1.86
Labor (9 Positions)	1.02	1.02	1.02
Taxes and Insurance	0.34	0.34	0.34
Total Operating Costs	12.55	20.55	28.55
Required Coal Market Price, $/MM Btu			
Utility Financing	0.84	1.32	1.80
Investor Financing	1.20	1.74	2.28

*Equivalent to a plant capital investment of $54.60/kw.

TABLE 5-31

Case 3, Deep Cleaned Fine Coal with 50% Meyers Process Bypass

Case product annual energy value, 43.4×10^6 MM Btu/yr	ROM coal cost		
	$10/Ton	$20/Ton	$30/Ton
Capital Related Requirements, \$MM			
Battery Limit Capital	13.26	13.26	13.26
Offsite Capital	9.13	9.13	9.13
Overhead and Profit	4.93	4.93	4.93
Engineering and Design	2.24	2.24	2.24
Contingency	4.43	4.43	4.43
Total Plant Investment*	33.99	33.99	33.99
Interest for Construction	5.74	5.74	5.74
Startup Costs	5.28	9.12	12.96
Working Capital (Utility Financing)	4.83	8.88	12.92
Working Capital (Investor Financing)	5.51	9.66	13.80
Total Capital Related Costs (Utility)	49.84	57.73	65.61
Total Capital Related Costs (Investor)	50.52	58.51	66.49
Operating Costs, \$MM/Yr			
Raw Material (Coal)	19.20	38.40	57.60
Chemicals (Lime, Scrap)	0.11	0.11	0.11
Supplies	0.69	0.69	0.69
Disposal	0.84	0.84	0.84
Utilities	2.62	2.62	2.62
Labor (14 Positions)	2.01	2.01	2.01
Taxes and Insurance	0.92	0.92	0.92
Total Operating Costs	26.39	45.59	64.79
Required Coal Market Price, \$/MM Btu			
Utility Financing	0.73	1.20	1.66
Investor Financing	1.11	1.63	2.15

*Equivalent to a plant capital investment of \$58.70/kw.

TABLE 5-32

Case 4, Deep Cleaned Coarse Coal with 50% Meyers Process Bypass

Case product annual energy value, 42.2 x 10^6 MM Btu/yr	ROM coal cost		
	$10/Ton	$20/Ton	$30/Ton
Capital Related Requirements, $MM			
Battery Limit Capital	4.20	4.20	4.20
Offsite Capital	6.50	6.50	6.50
Overhead and Profit	2.35	2.35	2.35
Engineering and Design	1.07	1.07	1.07
Contingency	2.11	2.11	2.11
Total Plant Investment*	16.23	16.23	16.23
Interest for Construction	2.74	2.74	2.74
Startup Costs	4.92	8.76	12.60
Working Capital (Utility Financing)	4.51	8.55	12.59
Working Capital (Investor Financing)	4.89	9.04	13.18
Total Capital Related Costs (Utility)	28.40	36.28	44.16
Total Capital Related Costs (Investor)	28.78	36.77	44.75
Operating Costs, $MM/Yr			
Raw Material (Coal)	19.20	38.40	57.60
Chemicals (Lime, Scrap)	0.39	0.39	0.39
Supplies	0.38	0.38	0.38
Disposal	0.96	0.96	0.96
Utilities	1.96	1.96	1.96
Labor (11 Positions)	1.29	1.29	1.29
Taxes and Insurance	0.44	0.44	0.44
Total Operating Costs	24.62	43.82	63.02
Required Coal Market Price $/MM Btu			
Utility Financing	0.66	1.14	1.62
Investor Financing	0.88	1.41	1.95

*Equivalent to a plant capital investment of $23.84/kw.

TABLE 5-33

Upgrading (Processing) Costs

	ROM Coal Cost		
	$10/Ton MM/Btu	$20/Ton MM/Btu	$30/Ton MM/Btu
Case 1			
Utility Financed	$0.43	$0.50	$0.57
Investor Financed	0.92	1.05	1.17
Case 2			
Utility Financed	0.43	0.51	0.58
Investor Financed	0.79	0.93	1.06
Case 3			
Utility Financed	0.32	0.39	0.44
Investor Financed	0.70	0.82	0.93
Case 4			
Utility Financed	0.25	0.33	0.40
Investor Financed	0.47	0.60	0.73

B. Dow Chemical — USA Design and Cost Estimation Studies

W.F. Nekervis and E.F. Hensley (36) performed a very detailed and creative study of the Meyers Process which resulted in a design of a commercial scale plant. They also prepared process cost estimations based on their design and studied variations from the baseline design directed toward cost reduction. Their design utilized the reaction rate data developed in TRW's first bench-scale study (13), presented in part in Section II.D.2. Their work is summarized in the four following sections together with some abstractions and paraphrasing from Reference 36.

1. PROCESS DESIGN

The feed and output coal specifications (Tables 5-34, 5-35 and 5-36) are based on the raw Lower Kittanning coal and the results obtained by Koutsoukos (13) in the initial Meyers Process bench-scale project.

The plant was assumed to be a turn-key stand-alone plant suitably located in Appalachia near a major branch of a coal hauling railroad. The plant would remove 95% of the pyrite from coal with an annual capacity of 3,000,000 metric tons of product coal/yr. Provision was made for removal of byproducts and wastes.

TABLE 5-34

Feed Coal Specification (35)

Lower Kittanning Coal, 200 mm (8 in.) top size	
Proximate Analysis:	%
Fixed Carbon	58.54
Volatiles	20.69
Ash	20.77
TOTAL	100.00
Heating Value, Cal/gm	6,745 ± 44
Heating Value, Btu/lb	12,140 ± 80
Moisture, assumed	10%
Rank: Medium Volatile Bituminous	
Sulfur Forms:	% w/w
Pyritic	3.58
Sulfate (Treat as Inert Solid)	0.04
Organic (Treat as Inert Solid)	0.67
TOTAL	4.29
Pyrite Content:	
(3.58% pyritic S) x (55.847 + 32.06 x 2)/	
(2 x 32.06) = 6.7% FeS_2	
Composition:	%
Coal (Excluding FeS_2)	93.3
Pyrites (as FeS_2)	6.7
TOTAL (Dry Basis)	100.0
Moisture (Assumed)	10.0
TOTAL	110.0

TABLE 5-35

Product Coal Specification (36)

Composition:	%
Coal (Excluding FeS_2)	99.64
Pyrites (as FeS_2)	0.36
TOTAL (Dry Basis)	100.00
Moisture	4.00
TOTAL	104.00

Size:

Compacted to 13 to 38 mm (1/2 in. to
1-1/2 in.) lumps with minimum of fines.

Detail of Pyrites Removed:

	Parts by Weight	
Composition:	Feed	Product
Coal	100.00	100.00
FeS_2	7.18	0.36
TOTAL	107.18	100.36

$(7.18 - 0.36)/7.18 = 95\%$ FeS_2 Removed

TABLE 5-36

Sulfur Forms in Product Coal (36)

Sulfur Form	% in Product
Unchanged Sulfur from Feed Coal Source:	
Organic	0.714
Sulfate	0.043
Sulfur or Compounds Left after Reactions and Washing:	
Iron Pyrites	0.185
Elemental Sulfur	0.027
Ferrous Sulfate	0.005
Ferric Sulfate	0.029
Sulfuric Acid	0.009
Subtotal from feed or reactions	1.012
Sulfur Compounds Added as Binder:	
Lignin Sulfonate	0.075
TOTAL based on dry processed coal	1.087

A block process flow sheet is shown in Figure 5-24.

The major differences from the Van Nice (16) process approach are: a) Section II − 100 mesh coal is used rather than 14 mesh top size, b) Section II − the reactor operates at $152^{\circ}C$ rather than $120^{\circ}C$, c) Section II − the reactor is a continuous up-flow tower with slurry and O_2 fed at the bottom rather than a horizontal continuous flow tank with oxygen fed into several cells, d) Section III − solvent (naptha) is used to remove elemental sulfur before water wash rather than sulfur vaporization after water wash, e) Section III − water is used to displace residual solvent, f) Section IV − product coal is compacted rather than fed directly into a power plant, and g) Section VI − product iron sulfate is removed by evaporation of part of the recycle solution to dryness, rather than crystallization and liming.

A full description of the flow diagram, equipment options and several detailed piping and instrumentation diagrams are presented by Nekervis and Hensley in Reference 36. The thermal efficiency was calculated to be 90% from run-of-mine coal to desulfurized and crushed product coal.

2. PROCESS ECONOMICS

A capital cost was estimated from a detailed equipment list for the stand-alone plant and full costs for manpower, utilities, materials, land etc., were developed. The complete operating cost summary is shown in Table 5-37 leading to a processing cost of $13/short ton (52¢/MM Btu) of product coal including binder and about 4% moisture as sold in 1973 dollars. Capital was depreciated on an investor financing basis for a 20.5% return on sales.

The effect of production rate (plant size) on processing cost is shown in Figure 5-25. The figure indicates that the treating cost is not greatly affected by production rate above about 3.0×10^6 tons/yr.

Various cost reduction cases differing from the basic case were also considered. In one typical example it was assumed that: a) the desulfurization plant is installed on the site of an existing complex that has coal handling facilities, utilities and services (such as at the power plant), and b) 14 mesh top size is used rather than 100 mesh. The total fixed capital cost (Table 5-38) is reduced by 2/3 and the cost of product coal is reduced to $7.32/ton. Other cost reduction alternatives suggested by Nekervis and Hensley include: a) use of faster

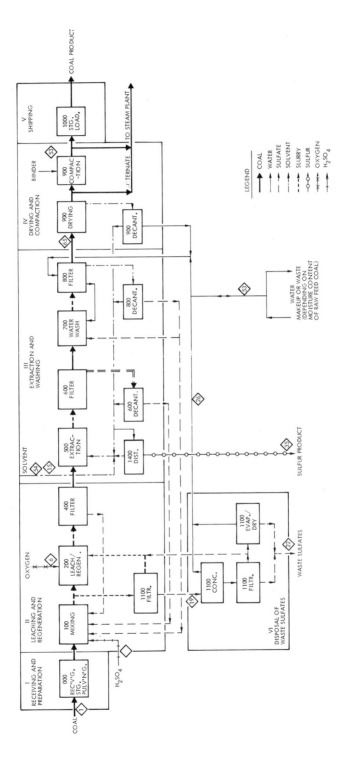

Figure 5-24. Dow-Midland Process Flow Diagram (36)

TABLE 5-37

Operating Cost Estimate Summary (36)

Annual production:	
Dry basis	$3,000 \times 10^3$ MT ($3,300 \times 10^3$ short T)
As sold, 1.5% binder and 4% moisture	$3,170 \times 10^3$ MT ($3,500 \times 10^3$ short T)
Working Capital:	$ 9,050 \times 10^3$
Total Capital:	$155,050 \times 10^3$

Cost Item		$M/yr	Dry basis $/short T	As sold $/short T
Raw Materials and by-products				
Coal				
Ash & coal loss	264×10^3	2,323	0.704	0.664
To fuel	297×10^3	2,614	0.792	0.747
To product	$3,300 \times 10^3$	29,040	8,800	8.297
Sulfuric acid		1,901	0.576	0.543
Naptha		178	0.054	0.051
Chemical agents		165	0.050	0.047
Binder		5,610	1.700	1.603
Sulfur credit		- 594	-0.180	-0.170
Nitrogen credit		- 323	-0.098	-0.092
Subtotal		40,914	12.398	11.690
Capital Related Costs				
Depreciation		12,043	3.649	3.441
Maintenance		5,840	1.770	1.669
Insurance and taxes		2,920	0.885	0.834
Subtotal		20,803	6.304	5.944
Other Costs				
Power		2,138	0.648	0.611
Water and waste disposal		2,478	0.751	0.708
Labor and miscellaneous		6,885	2.086	1.967
General and administrative		270	0.082	0.077
Subtotal		11,771	3.567	3.363
Total and unit cost		73,488	22.269	20.997
Treating cost		44,448	13.469	12.700
			$/MT	$/MT
Total unit cost, metric		73,488	24.496	23.182
Treating cost, metric		44,448	14.816	14.022

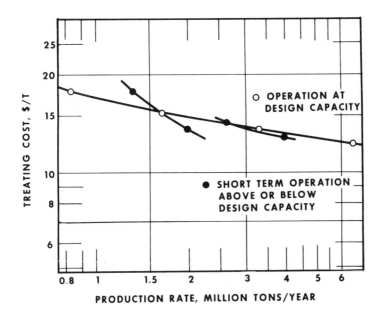

Figure 5-25. Production Rate vs Treating Cost (34)

reacting coals (see Section II.B.), b) control leach solution acidity to reduce elemental sulfur make and use of a solvent with very high sulfur capacity.

3. PROCESS BY-PRODUCTS

The value and suggested disposition of by-products of the Meyers Process are shown in Table 5-39. Clearly, a single 420 ton/yr Meyers Process plant, treating a 3.2% pyritic sulfur coal at 95% removal, would saturate the U.S. iron sulfate market forcing disposal (by liming or in lined pits) or conversion to iron ore grade material. Elemental sulfur would not necessarily be a problem, since it can be sold or easily stored.

TABLE 5-38

Dow-Midland Cost Reduction Case

Capital Reduction, $M		
Process Capital		
Leaching	15,000	
Extraction	2,500	
Waste Sulfates	6,500	
Subtotal, Process	24,000	
Other Capital		
Site, Buildings, General	3,400	
Utilities		
Waste Sulfates	4,490	
Remainder	2,520	
Allowances		
Process and Other Capital Reductions	2,750	
Subtotal, Other	13,160	
Total Fixed Capital Reductions	37,160	
Ratio to Base Case Chemical Processing (117.298M)		0.32
Operating Costs Reduction, $M/yr		
Capital Related		
Labor and G&A	1,360	
Other Capital Related	5,550	
Unit Ratio Based		
Utilities, Waste Disposal, Coal	1,500	
Equivalent to Fuel & Ash Loss, Less		
Reduced Sulfur Credit		
TOTAL, Operating Cost Reduction	8,460	
Effect on Chemical Processing Regions (8460 ÷ 3300)		-$2.56/T
Alternative Unit Cost		
Base Case Chemical Processing		$9.88/T
Less Reductions		2.56
Reduced Total		$7.32/T

TABLE 5-39

Value and Disposition of By-products from Coal Leaching (36)

By-product	By-product production for plants treating 100 and 420 T of coal/hr			U.S. Market		Most likely disposition for by-product from numerous plants
	100	420				
	T/hr	T/hr	1,000 Tons/Yr	1,000 T/yr	$/T	
Ferric Sulfate	3.7	15.5	122	100	37	Haul to an abandoned mine or quarry for disposal at $2.50/T or build a processing unit to upgrade to iron oxide if the process economics and markets warrant it.
Ferrous Sulfate	4.3	18.1	142	112	24	
Iron Oxide: Pigment Grade	6.0	25.2	198	140	300	Small amount could go into pigment market.
Iron Ore Grade				45,000	10	Entire output could be shipped to steel mills within 100 mile radius if the quality was acceptable.
Sulfur, Elemental	1.2	5.0	39	10,500	12 to 32	Sulfur should be marketable if there are customers within a 100 mile radius. Could also be collected at a central river-front location and shipped by barge.

C. Exxon Research and Engineering Co. Design and
Pollution Control Studies

E.M. Magee (37) of Exxon performed a detailed evaluation of the Meyers
Process from the standpoint of its potential for affecting the environment. Thermal efficiency was also evaluated. A process flow diagram was developed based
on the data in the initial TRW report (13) but revised for assessment of pollution
and thermal efficiency.

A modified process design was developed by Magee with the following
major changes over the Dow approach: a) feed coal pyritic sulfur content was
assumed to be 3.21% (dry basis) rather than 3.58%, b) feed coal is reacted at
80% less than 200 mesh to enhance rates and provide a product suitable for
direct firing (without compaction) in a utility boiler, and c) coal drying is
eliminated.

The modified flow diagram is shown in Figure 5-26. An important energy
savings arises in Magee's use of 600 psig steam to drive the oxygen plant and
extraction steam at 115 psig for the process area. Magee calculated the thermal
efficiency of the Meyers Process using his modified design at 92.1%.

The mass balance, by process streams (Figure 5-26) was calculated (Table
5-40) for the purpose of analyzing possible pollution sources.

An analysis for possible polluting effluents revealed the following: 1)
essentially no air effluents are emitted from the reaction, product drying or
iron sulfate recovery sections, 2) the sulfur removal section is completely enclosed, except for vents, allowing no effluents to the air, 3) possible dusting
from the solid iron sulfate and elemental sulfur areas is handled by keeping the
iron sulfate moist and the sulfur as a liquid or in a closed storage system. Moisture containing air from the cooling tower represents the largest effluent to the
atmosphere. Magee concluded that in this particular plant the cooling water
should be relatively free of volatile materials; pressures on the heat exchangers
are low and, except for the organic solvent, no volatile organics have been
reported as being present in the reaction system. Fog formation can sometimes present a problem with cooling towers. The extent of this problem, however, is determined in large part by the plant location. Drift loss from the cooling tower can cause dust problems when the solids in the cooling water are
deposited. Magee also expected that cooling tower blowdown would be sent
to an evaporation pond. Due to the nature of the present process, Magee felt
that there should be no noxious fumes from this pond if there are no leaks in the
naptha heat exchangers.

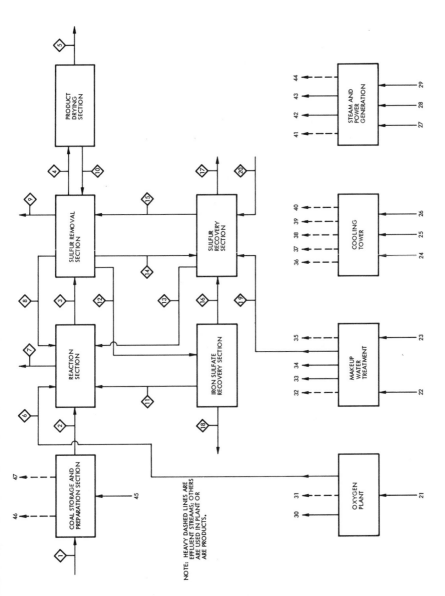

Figure 5-26. Flow Diagram — Modified Meyers Process (37)

TABLE 5-40

Stream Identification for Modified Process, Lb/Hr

Stream	1	2	3	4	5	6	7	8	9	10
Coal*	188,000	188,000	188,000	187,800	187,800					
FeS_2	12,000	12,000	602	602	602					
S			2,438							
$FeSO_4$			30,060					18,000		
$Fe_2(SO_4)_3$			122,798	200	200			73,800		
H_2SO_4			6,800					4,000		
H_2O	20,000	20,000	662,800	47,000	37,600		660	466,400	200	
Solvent				9,400					200	9,400
O_2						7,296				
Inert						40	40			
TOTAL	220,000	220,000	1,013,498	245,020	226,220	7,336	700	562,200	400	18,000
				245,002**	226,202**					

*Feed coal, without pyrites and moisture.
**Apparent mathematical error.

TABLE 5-40 (Continued)

Stream Identification for Modified Process, Lb/Hr

Stream	11	12	13	14	15	16	17	18	19	20
Coal		200						200		
FeS_2										
S				2,438			2,438			
$FeSO_4$	1,000	9,660	2,400	2,400				8,660		
$Fe_2(SO_4)_3$	31,600	38,998	9,800	9,800				7,398		
H_2SO_4	2,200	2,200	600	600						
H_2O	126,000	231,000	50,460	65,400	137,800	104,400			18,460	
Solvent				255,800	256,000					
O_2										200
Inert										
TOTAL	161,400	282,058	63,260	336,438	393,800	104,400	2,438	16,258	18,460	200

TABLE 5-40 (Continued)

Stream Identification for Modified Process, Lb/Hr

21 – Air to O_2 Plant	31,520	
22 – Chemicals	–	
23 – Water	153,850	
24 – Air	12,700,000 (4×10^9 scfd)	Air to cooling tower
25 – Water	135,560	Makeup water to cooling tower at 85°F
26 – Water	7,083,600	Water plus makeup recirc. to cooling tower at 105°
27 – Product Coal	13,234 (Dry)	Product coal to utility boiler
28 – Air	141,229	Air to boiler
29 – Water	126,000	Boiler feed water (includes 6000 lb/hr makeup water)
30 – Water	170	Moisture from air to boiler feed water
*31 – Nitrogen	24,050	Vent from O_2 plant, can be used in coal silos
*32 – Sludge	–	From treating makeup water
33 – Boiler Feed Water Makeup	5,830	To steam generation
34 – Cooling Water	129,560	Makeup
*35 – Backwash	–	
*36 – Air From Cooling Tower	12,700,000 (4×10^9 scfd)	Air from cooling tower
*37 – Drift Loss Water	14,160	Water mist to air
*38 – Water Vapor	100,000	Water evaporated from cooling tower
*39 – Blowdown Water	21,400	Purge from cooling tower to holding pond
40 – Cooling Water	7,083,600	–
*41 – Flue Gas	154,570	From utility boiler
42 – Blow Down Water	6,000	To cooling tower makeup
43 – Steam	120,000	
*44 – Ash	2,541	
45 – Rain**	e.g. 6" Rain in 24 Hours	Rain on coal storage and preparation area
*46 – Rain Run Off**	e.g. 6" Rain in 24 Hours	From coal storage and preparation area
*47 – Dust	–	From coal preparation collect in bag filters

*These streams are emitted to the environment.

**Not applicable if all storage is in silos.

TABLE 5-41

Power Requirements

Utilities	Electricity, kw
Consumed	
Coal Preparation	2,740
Steam to vaporize H_2O in plant	—
Oxygen Plant	—
Power generation	—
Cooling water pumps	420
Cooling tower fans	260
Boiler feed water pumps	110
Rest of plant	1,000
	4,520

All other plant areas were similarly assessed with the conclusion that the only pollution potential arises from the disposition of product iron sulfate solid, together with precipitated trace elements and possible dusting from solid sulfur.

Thus, both the iron sulfate and sulfur products would have to be stored under suitable conditions, e.g., the iron sulfate in lined pits and the sulfur away from the wind, possibly in a landfill.

An electricity balance is shown in Table 5-41 for a coal desulfurization plant of a size to supply a 250 MW baseload power plant. The electrical load for the desulfurization unit is 4.5 MW which is less than 2% of the electrical output of the power plant.

REFERENCES

1. R.A. Meyers, J.W. Hamersma, J.S. Land and M.L. Kraft, Science, 117: 1187 (1972).

2. R.A. Meyers, "Removal of Pyritic Sulfur from Coal Using Solutions Containing Ferric Ions," U.S. Patent 3768988 (1973).

3. B.K. Mazumdar, Fuel, 41: 121 (1962).

4. L. Lorenzi, Jr., J.S. Land, L.J. Van Nice, E.P. Koutsoukos and R.A. Meyers, Coal Age, 77: 76 (1972).

5. R.A. Meyers, "Desulfurization of Coal," paper presented at the Symposium on the Desulfurization of Coal, 71st National Meeting of the American Institute of Chemical Engineers, Dallas, Texas, Feb. 22, 1972.

6. R.A. Meyers, "Removal of Pollutants from Coal," paper presented at the Symposium on Coal Conversion and Environment, American Geophysical Union, Washington, D.C., April 19, 1972.

7. J.W. Hamersma, M.L. Kraft, W.P. Kendrick and R.A. Meyers, "Chemical Desulfurization of Coal to Meet Air Pollution Control Standards," Preprints Div. of Fuel Chemistry, Am. Chem. Soc., 19(2) 33 (1974).

8. L. Lorenzi, Jr., L.J. Van Nice, M.J. Santy and R.A. Meyers, "Plant Design of a Method for Chemical Desulfurization of Coal," Preprints Div. of Fuel Chemistry, Am. Chem. Soc., 19(2) 43 (1974).

9. E.J. Sercombe, J.K. Gary, Brit. Patent 1,143,139 (1969).

10. V.V. Emilov, Y.P. Romanteev, Y.P. Shchurouskii, Ya, Tr. Inst. Met. Obogashch, Akad. Nauk Kaz. SSR, 30: 55-64 (1969).

11. L. Liepna, B. Macejevskis, Dokl. Akad, Nauk SSR, 173: 1336-1338 (1967).

12. J.W. Hamersma, M.L. Kraft, E.P. Koutsoukos and R.A. Meyers, in Advances in Chemistry Series, No. 127, American Chemical Society, Washington, D.C., 1973.

13. J.W. Hamersma, E.P. Koutsoukos, M.L. Kraft, R.A. Meyers, G. J. Ogle and L.J. Van Nice, "Chemical Desulfurization of Coal: Report of Bench-Scale Developments," EPA R2-73-173, U.S. Environmental Protection Agency, Washington, D.C., 1973.

14. G. Nickless, ed., Inorganic Sulfur Chemistry, Elsevier Publ. Co., New York, 1968, p. 95.

15. J.W. Hamersma, M.L. Kraft, C.A. Flegal, A.A. Lee and R.A. Meyers, "Applicability of the Meyers Process for Chemical Desulfurization of Coal: Initial Survey of Fifteen Coals," EPA-650/2-74-025, U.S. Environmental Protection Agency, Washington, D.C., 1974.

16. E.P. Koutsoukos, M.L. Kraft, R.A. Orsini, R.A. Meyers, M.J. Santy and L.J. Van Nice, "Final Report Program for Bench-Scale Development of Processes for the Chemical Extraction of Sulfur from Coal," EPA 600/2-76-143a, U.S. Environmental Protection Agency, Washington, D.C., May 1976.

17. J.R. Pound, J. Phys. Chem., 43: 955 (1939).

18. F.J. Bartholomew, Chem. Eng. 57: (8) 118 (1950).

19. T. Fujimori and K. Ishikawa, Fuel, 51: 120 (1972).

20. C.T. Mathews and R.G. Robins, Australas. Inst. Mining Met. Proc., No. 242: 47-56 (1972).

21. C.G. Maier, "The Ferric Sulfate-Sulfuric Acid Process," U.S. Dept. of Commerce, Bureau of Mines Bull. No. 260, 1972.

22. J.W. Mellors, "A Comprehensive Treatise on Inorganic and Theoretical Chemistry," John Wiley and Sons, New York, Vol. 10: 1961, pp 89-90.

23. W. Kunda, B. Rudyk and V.N Nackiw, Can. Min. and Met. Bulletin, 810, July 1968.

24. P.C. Singer and W. Stumm, "Oxidation of Ferrous Iron: The Rate-Determining Step in the Formation of Acidic Mine Drainage," Water Pollution Control Research Series, DAST 28, 14010, U.S. Dept. of the Interior, June 1969.

25. E.P. Koutsoukos, R.A. Orsini, G.J. Ogle and R.A. Meyers, "Chemical Desulfurization of Coal by the Meyers Process," 73rd National Meeting of the American Institute of Chemical Engineers, Salt Lake City, Utah, August 13, 1974.

26. J.W. Hamersma and R.A. Meyers, unpublished results.

27. R.A. Meyers, J.W. Hamersma, R.M. Baldwin, J.G. Handwerk, J.H. Gary and J.O. Golden, Preprints Div. of Fuel Chemistry, Am. Chem. Soc., 20: (1) 234 (1975).

28. H. Mark, ed., Kirk-Othmer Encyclopedia of Chemical Technology, Vol. 19, Interscience Publ., New York, 1969, p. 342.

29. F.P. Haver and R.D. Baker, "Improvements in Ferric Chloride Leaching of Chalcopyrite Concentrate," Bureau of Mines Report of Investigations RI 8007, Dept. of Interior, Washington, 1975.

30. R.A. Orsini, unpublished results (May 1976).

31. A. Watanbe, Kagaku Kogaku, 31: (7) 676 (1967).

32. F.J. Bartholomew, J. Chem. Eng., 57: (8) 118 (1950).

33. V. Kramarsic and S. Cernivec, "Method of Regenerating Spent Liquors," Yugoslavia Pat. Spec. 6-13-58.

34. J.W. Hamersma and M.L. Kraft, "Applicability of the Meyers Process for Chemical Desulfurization of Coal: Survey of 35 Coal Mines," EPA-650/2-74-025a, U.S. Environmental Protection Agency, Washington, D.C., 1975.

35. U.S. Environmental Protection Agency, Federal Register, 36: 24876 (1973).

36. W.F. Nekervis and E.F. Hensley (Dow Chemical, U.S.A.), "Conceptual Design of a Commercial Scale Plant for Chemical Desulfurization of Coal," Environmental Protection Technology Series, EPA-600/2-75-051, U.S. Environmental Protection Agency, Washington, D.C., September 1975.

37. E.M. Magee (Exxon Research and Engineering Co.), "Evaluation of Pollution Control in Fossil Fuel Conversion Processes, Coal Treatment: Section 1, Meyers Process," Environmental Protection Technology Series, EPA-650/2-74-009k, U.S. Environmental Protection Agency, Washington, D.C., September 1975.

38. M. Rasin Tek (U. of Michigan), "Coal Beneficiation," in Evaluation of Coal Conversion Processes, PB-234202, available through National Technical Information Service, Springfield, Va., 1974.

39. E.F. Hensley (Dow Chemical, U.S.A.), "Pyritic Sulfur Removal from Coal," presented at the 30th Annual Am. Chem. Soc. Fall Scientific Meeting, Midland, Michigan, November 1973.

40. G.C. Sinke (Dow Chemical, U.S.A.), "The Desulfurization of Coal," presented at the 30th Annual Am. Chem. Soc. Fall Scientific Meeting, Midland, Michigan, November 1973.

41. W.F. Nekervis (Dow Chemical, U.S.A.), "The Economics of Desulfurization of Coal," presented at the Annual Am. Chem. Soc. Fall Scientific Meeting, Midland, Michigan, November 1973.

42. J.W. Leonard and D.R. Mitchell, eds., Coal Preparation, Am. Inst. of Mining, Met. and Petrol. Engs. Inc., New York, 1968.

CHAPTER 6

PYRITIC SULFUR REMOVAL PROCESSES — OXYGEN IN AQUEOUS SOLUTION

I.   Atmospheric Pressure Methods

II. Elevated Pressure Methods
     A. Process Chemistry and Data
     B. Process Design

III. Bacteria Catalyzed Methods

The oxidation of coal with air or oxygen at temperatures below the coal ignition temperature has been extensively studied (1). Rates of oxidation of the organic coal matrix have been determined at temperatures as low as $70^{\circ}C$, wherein peroxide complexes are formed initially, followed by more stable combinations of carbon and oxygen such as phenol and carbonyl. High temperatures result in decomposition to carbon monoxide, carbon dioxide and water with rapid weight loss beginning about $200^{\circ}C$.

Coal rank was shown to play an important role in the degree of oxidation of coal, with the higher rank coals generally undergoing less oxidation (see Chapter 5, Sections II and IV). The results of Fe^{+3} leaching survey studies (see Chapter 5, Section IV) show that the oxidizability of Appalachian coal is lowest, increasing for Interior Basin coals, and is quite high for Western coals. Thus, rank is indeed an important factor in the stability of coal toward oxidizing agents.

Soluble iron sulfate naturally present in coal quickly dissolves into aqueous solution and in the presence of oxygen may become the chain carrier for the oxidation of pyrite by air. This has been shown in acid mine drainage mechanistic studies (Chapter 3, Section I.B.) and indicated by the results of simultaneous pyrite leach and Fe^{+3} regeneration studies (Chapter 5, Section II.D.). Hence, the oxidation of pyrite in coal using an aqueous oxygen or air treatment may well be a Fe^{+3} oxidation process, essentially identical to the simultaneous pyrite leach and ferric sulfate regeneration studies (Meyers Process) discussed in Chapter 5. Pyritic sulfur removal studies using aqueous oxygen are discussed in the following three sections on atmospheric pressure methods, elevated pressure methods and bacterial catalyzed oxygen treatment.

I. ATMOSPHERIC PRESSURE METHODS

An early study of the air oxidation of pyrite in coal was performed by Powell and Parr (2) who found that 25% of the pyritic sulfur in an Illinois

No. 6 coal could be oxidized in a two-year period at room temperature in a tightly stoppered flask. When a slow current of air was passed through a flask containing an undried sample of Vermillion County Illinois coal, only about 10% of the coal pyrite was oxidized under the most favorable conditions of air supply, temperature and fineness of coal.

Li and Parr (3) later showed in more quantitative studies that 30-74% of the pyritic sulfur (15-24% of the total sulfur) of various Illinois coals (Figure 6-1) could be oxidized in a period of six weeks at $100^{\circ}C$ in a simple stationary bed of -80 mesh coal over which saturated oxygen was passed. The relative rates of oxidation of (-325 mesh) pyrite and marcasite (Figure 6-2) were also investigated under the conditions cited above. Marcasite had been reported (3) to be present in coal and to react at a rate differing from that of pyrite. The

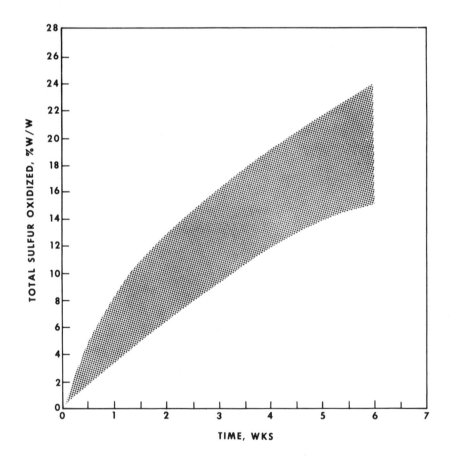

Figure 6-1. Reaction of Four Illinois Coal with Oxygen Saturated
with Moisture – $100^{\circ}C$

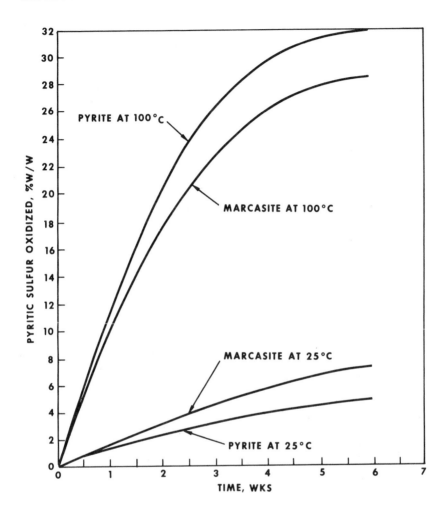

Figure 6-2. Low Temperature Reaction of Pyrite with Oxygen
Saturated with Moisture

results, however, indicated that there was no significant difference in reactivity. The minerals were found to oxidize at rates equivalent to that found for pyrite in the four coals (ground to -80 mesh) investigated (Figure 6-1). Apparently the pyrite in the coals had an effective particle size distribution roughly corresponding to the -325 mesh mineral pyrite. Moisture was shown (Figure 6-3) to be an important factor in the oxidation rate. The radical chain mechanism involving Fe^{+3} reactant, which must be inoperative without moisture, would account for the almost total suppression of oxidation found for anhydrous conditions, (see Chapter 3-I.B.2.c. for a discussion of the rate of direct reaction of O_2 with pyrite under anhydrous conditions).

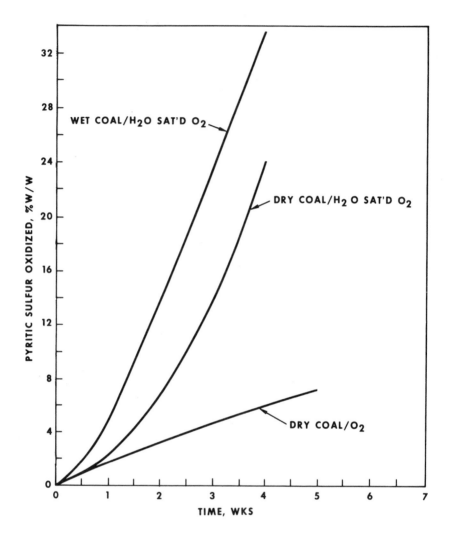

Figure 6-3. Reaction of Vermillion Coal Pyrite with
Oxygen – 100°C

Nelson, Snow and Keyes (4) extended the Li and Parr approach and improved the removal of pyritic sulfur by suspending the coal in water and passing fine air or oxygen bubbles through the suspension. Residual sulfate was removed from the coal by washing with hydrochloric acid. The coal was then dried at 104°C under vacuum. More specifically, an Illinois coal was pulverized to -60 mesh, suspended in an aqueous solution at temperatures up to 90°C, and air or oxygen was passed through at a very rapid rate (0.113 M^3/min). Indeed, the rate of oxidation was significantly improved; however, after one week of this treatment, a maximum of only 79% of the pyritic sulfur was oxidized. When ferric sulfate was added as a "catalyst," a slight rate improvement was observed (Table 6-1). Nelson, *et al* concluded that "the addition of ferric sulfate hastened the oxidation process to some extent but not as much as might be expected." Indeed, only 48% of the pyritic sulfur was removed in 24 hrs and the rate of removal had slowed considerably.

TABLE 6-1

Effect of Ferric Sulfate* on Air Oxidation of Pyritic Sulfur in Coal

Temp °C	Time hours	Pyritic sulfur, % w/w			Catalyst
		In original coal	In oxidized sample	Oxidized	
70	2	2.48	1.90	24.8	None
70	2	2.48	2.00	20.0	None
70	5	2.48	1.83	26.3	None
70	24	2.48	1.61	35.1	None
70	3	2.48	1.75	29.4	0.05% Fe^{+++} as $Fe_2(SO_4)_3$
70	24	2.48	1.55	37.5	0.05% Fe^{+++} as $Fe_2(SO_4)_3$
70	24	2.57	1.42	44.8	0.1% Fe^{+++} as $Fe_2(SO_4)_3$
70	24	2.57	1.38	46.3	0.1% Fe^{+++} as $Fe_2(SO_4)_3$
70	24	2.48	1.30	47.5	0.1% Fe^{+++} as $Fe_2(SO_4)_3$

*25g coal/4.5 liters of solution, 24 hrs at 70°C.

The increase in pyritic sulfur removal on addition of ferric sulfate is explainable by consideration of the demonstrated direct reaction of Fe^{+3} with coal pyrite (Meyers Process). Nelson, *et al* suspended the coal in 4.5 liters of 0.1% w/w Fe^{+3} solution/25g of 2.6% pyritic sulfur coal. This was enough Fe^{+3} to oxidize about 90% of the contained pyrite by simple direct oxidation to form ferrous sulfate and elemental sulfur (*cf* Section II). Although the authors were aware of the potential for Fe^{+3} to directly react with pyrite and form elemental sulfur as one of the products, this possibility was not considered in their paper. Further, the drying procedure used was adequate to vaporize the sulfur formed (*cf* Section III) so that its presence was not noted.

This line of investigation for development of a process for removal of pyritic sulfur from coal was not further pursued probably due to the low pyritic sulfur removal observed and the belief that hydrochloric acid was needed to remove generated iron sulfate. Obviously, the use of hydrochloric acid would not lend itself to process development, since it would be necessary to separate generated sulfate from chloride solution, a costly and difficult task. Moreover, the chloride ion would contaminate the coal.

Meyers and Hamersma (5) recently investigated the above "catalyzed" aeration technique utilizing identical conditions. Two Illinois coals (Orient No. 6 and Eagle mines) and two Appalachian coals (Isabella and Bird No. 3 mines) were treated (Table 6-2). It was concluded that the added ferric sulfate "catalyst" was the active agent as the presence of air only slightly increased the degree of pyritic sulfur removal over Fe^{+3} alone, and this for only three of four coals. Elemental sulfur was generated during the process whether or not air was present indicating that a common mechanism was operable, that of Fe^{+3} leaching with some regeneration of Fe^{+3} during aeration as previously noted for more concentrated Fe^{+3} solutions under similar conditions (cf – Chapter 5-II.D.).

TABLE 6-2

Effect of Ferric Sulfate and Air on Oxidation of Pyrite in Coal

Coal	Treatment*	Pyritic sulfur removal
Orient No. 6	$Fe_2(SO_4)_3$/Air	69%
	$Fe_2(SO_4)_3$	74%
Eagle	$Fe_2(SO_4)_3$/Air	66%
	$Fe_2(SO_4)_3$	50%
Isabella	$Fe_2(SO_4)_3$/Air	78%
	$Fe_2(SO_4)_3$	61%
Bird No. 3	$Fe_2(SO_4)_3$/Air	57%
	$Fe_2(SO_4)_3$	40%

*25g coal/4.5ℓ of solution, 0.1% w/w Fe^{+3} as $Fe_2(SO_4)_3$, 24 hrs at 70 ± 5°C.

II. ELEVATED PRESSURE METHODS

Ledgemont Laboratories performed a study of the leaching of pyritic sulfur in aqueous coal slurry at temperatures of 80-130°C and oxygen partial pressures of 100-300 psi. Workers at the Energy Research and Development Administration have recently studied the leaching of pyritic sulfur in aqueous coal slurry at 150-180°C pressurized with air to about 1000 psi ($P_{O_2} < 200$ psi). Experimental results, a process engineering description and capital costs are presented in the following sections.

A. Process Chemistry and Data

The Ledgemont experimentation (6) was conducted in a batch mode in high pressure autoclaves equipped with baffles and an agitator. Coal was slurried in distilled water to give a solids density of 20% w/w. No information is given as to the workup of the resulting desulfurized coal, although it can be assumed that the product coal was separated by filtration or centrifugation from the leach solution and then washed and dried at temperatures sufficient to prepare the coal for analysis. No mention is made of any attempt to isolate elemental sulfur, a likely process product, which would tend to volatilize along with water during most normal drying procedures.

The Ledgemont workers (6) believed that the reaction for dissolution of pyrite proceeded entirely according to Eq. 1, and that no elemental sulfur was formed because "starting with a neutralized solution, in our case avoided the formation of elemental sulfur." However, Ichikuni (7) had showed that oxygen treatment of mineral pyrite in aqueous solution forms only soluble sulfate at pH above 2.5 but forms elemental sulfur below pH of 2.5. The Ledgemont workers indicated that the final pH of the coal leach solution is about 1.4 after removal of about 90% of the pyritic sulfur. Clearly, a good portion of the pyrite dissolution during the oxygen leach must take place at a pH below the 2.5 level, and therefore significant amounts of elemental sulfur should be indeed expected, contrary to the conclusions of the Ledgemont authors.

$$FeS_2 + 3.5\ O_2 + H_2O \rightarrow Fe^{+2} + 2\ SO_4^{=} + 2H^{+} \qquad (1)$$

Curves for pyritic sulfur removal at two oxygen pressures are shown in Figure 6-4 which indicate that 90% removal of pyritic sulfur is obtained after 180 minutes of reaction at 130°C and pressures of 100-300 psi of oxygen. These results are quite comparable to those previously reported by Koutsoukos (8) and described in Chapter 5-II.D. See Chapter V, Figure 5-9 which shows approximately 70% removal of pyritic sulfur in 90 minutes at 120°C and 85 psi of oxygen, which is comparable to the removal reported in Figure 6-1 by Ledgemont at 100 psi of oxygen and 130°C in 70 minutes. Interestingly, these results were obtained on coals from different coal basins, the Lower Kittanning seam from the Appalachian basin (8) and the Illinois No. 6 coal from the Eastern Interior basin (6). It is informative to note that Koutsoukos, who performed careful material balanced experimentation for the isolation of product elemental sulfur, found the same amount of elemental sulfur product as normally found for ferric sulfate leaching even under these "no added iron" conditions. Thus, it would appear that the Ledgemont results were definitely comparable to the previous work with the exception of the discrepancy in elemental sulfur.

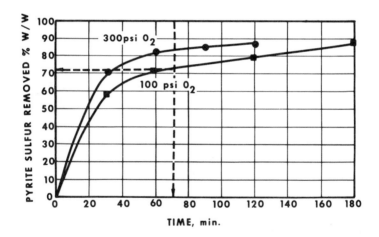

Figure 6-4. Ledgemont Process – Effect of Pressure on Pyrite Removal from Illinois #6 Coal at 130°C (6)

McKay and Halpern further noted that elemental sulfur is always formed, except for cases where no sulfuric acid is initially present, although in all cases the final pH was lowered to about 1. Further, it is likely that sulfur was formed even in this latter case but was overlooked due to the low pyrite conversion obtained at low acid concentration (i.e., <20% pyrite reaction). The small amount of sulfur which was probably present, could not be measured by the relatively insensitive analytical technique used (i.e., weight difference after aniline extraction).

The effect of coal particle size is shown in Figure 6-5. As indicated in the figure, 100 mesh top-size reacts significantly faster than the 40 mesh. This is in concert with the results for simultaneous leaching and regeneration in the ferric sulfate/oxygen leaching system discussed in Chapter 5. The effect of temperature is shown by the curves in Figure 6-6. An activation energy of 14 kcal/mole utilizing an Arrhenius type equation was used to generate these curves, which is similar to that reported by McKay and Halpern (9) for the leaching of mineral pyrite at these temperatures. The rate of sulfur removal as a function of pressure, calculated from the data in Figure 6-4, shows a square root dependence on the oxygen pressure which the authors point out was noted by Warren (10), but they failed to note that the results of McKay and Halpern (9) showed a first order of dependence on the partial pressure of oxygen.

Finally, the authors found that significant combustion of the Illinois No. 6 coal occurs under the conditions utilized, giving rise to carbon dioxide and smaller amounts of carbon monoxide and hydrocarbons. This secondary reaction is highly dependent on oxygen partial pressure (Figure 6-7), consuming up to 6% of the weight of coal in oxygen at 300 psi oxygen partial pressure. Thus, as shown in Chapter 5, Figure 5-8, optimal pressure leaching of pyrite from coal occurs at lower oxygen partial pressure and temperatures than those utilized in this study. However, these results are for Illinois No. 6 coal which is characteristically more susceptible to oxidation than the Appalachian coals (see Chapter 5, Section IV) so that there would be less coal reaction with oxygen, under the conditions utilized by the Ledgemont researchers, with an Appalachian coal.

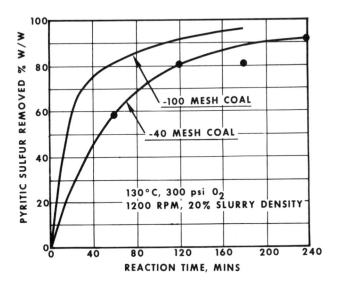

Figure 6-5. Effect of Coal Top-Size on Pyritic Sulfur Removal

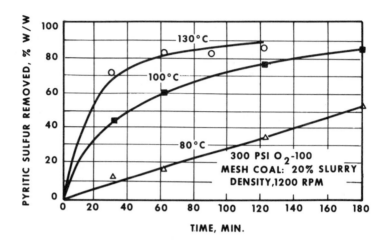

Figure 6-6. Effect of Temperature on Pyrite Removal from
Illinois #6 Coal (6)

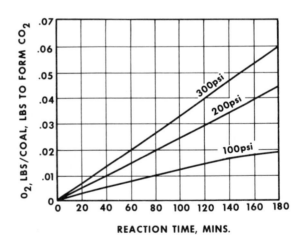

Figure 6-7. Effect of Pressure on Oxidation of Illinois #6
Coal to CO_2 at 130°C (6)

Friedman, Lacount and Warzinski (11) obtained results similar to those
of Ledgemont (6) and Koutsoukos (8) using air at a total pressure of 1000 psi
in place of oxygen and using temperatures of 150-180°C (Table 6-3). The partial pressure of oxygen at these temperatures would be slightly less than 200 psi.
The reported removal of greater than 90% pyrite in one hour is in line with the
results shown in Figure 6-4 obtained by the Ledgemont workers (70-80% removal

in one hour on 100 mesh coal at 130°C and 100-300 psi of oxygen) in view of the fact that Friedman, Lacount and Warzinski performed their experimentation on 200 mesh x 0 coal while Ledgemont used 100 mesh x 0.

TABLE 6-3

Pyrite Leaching* by Air-Steam Process (11)

Coal (seams)	T, °C	Pyritic sulfur content, % w/w	
		Initial	Treated
Illinois No. 5	150	0.9	0.1
Minshall	150	4.2	0.2
Lovilia No. 4	150	4.0	0.3
Pittsburgh	160	2.8	0.2
Lower Freeport	160	2.4	0.1
Brookville	180	3.1	0.1

*Nominal conditions: 1000 psi total pressure, 1 hr leach time.

B. Process Design

The Ledgemont workers (6) prepared a process schematic for oxygen leaching of coal at temperatures of 100-130°C and pressures up to 300 psi (Figure 6-8). The system is almost identical to that reported for the Meyers Process in Chapter 5 with the exception of the need to completely neutralize the leach solution so as to obtain what amounts to distilled water, and the elimination of any step for the removal of generated elemental sulfur. In Chapter 5, however, it was shown that an elemental sulfur cleanup step is almost certainly needed, and in that case the difference between the Meyers Process and the Ledgemont Process simply reduces to the relative concentration of iron in the leach solution and the method of removal of generated iron sulfate prior to recycle of leach solution. The capital cost estimate (12) for a plant based on the schematic in Figure 6-8, indicates that the plant cost would be about $35/annual ton of coal (Table 6-4) treated or roughly $110/KW utility name plate capacity for a desulfurization plant sized at 8000 tons/day of feed, overlooking coal loss due to reaction with oxygen (see Figure 6-7) and coal used for process heat. Also overlooked is the need for an elemental sulfur removal section.

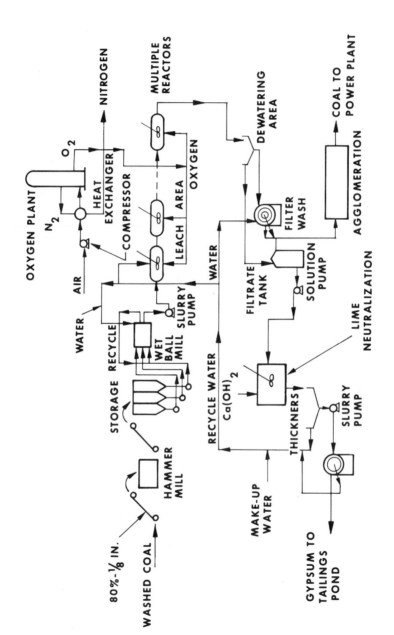

Figure 6-8. Flowsheet for Ledgemont Process for Leaching
of Pyritic Sulfur (6)

TABLE 6-4

Ledgemont Process Capital Cost Estimate* (12)

Section	Cost** $/ton/hr
Coal handling and cleaning	10,900
Feed bins and conveyor	2,300
Crushing and milling	12,700
Reactors	35,600
Oxidant	40,200
Washing (filters and thickeners)	20,000
Agglomeration	25,600
Neutralization	14,700
	162,000
Contingency (25%)	40,500
Fixed capital, $/ton/hr	202,500
Working capital, $/ton/hr	39,000
Land, start-up, interest during construction, $/ton/hr	42,500
Total investment, $/ton/hr	284,000
$/annual ton	35.4

*Plant capacity of 8000 tpd.

**1975 dollars.

III. BACTERIA CATALYZED METHODS

It has long been known that certain iron and sulfur oxidizing bacteria found in acid mine waters are important in the oxidation of pyrite present in coal (see Chapter 3, B.2.c.). Zarubina (13) studied the oxidation of pyrite in coal by oxygen in the presence of the bacterium *Thiobacillus ferrooxidans* and reported that the organism catalyzed the removal of 23-27% of the pyrite of a Soviet coal in a period of 30 days.

Silverman, Rogoff and Wender (14) found that the sulfur oxidizing bacterium *Thiobacillus ferrooxidans*, usually found in acid mine waters, was unable to attack pyritic sulfur in coal, while the iron oxidizing bacterium *Ferrobacillus ferrooxidans*, also found in acid mine waters, readily catalyzed the oxidation of pyrite in coal. The rate of pyritic sulfur removal is shown in Figure 6-9, where it can be seen that up to about 60% pyritic sulfur removal can be obtained in a period of 8 days at room temperature. It can also be seen that particle size has an effect somewhat comparable to that found for higher pressure and temperature oxygen dissolution of pyrite. The ability of this bacterial system to desulfurize various classes of coals is shown in Table 6-5, which indicates

TABLE 6-5 (14)

Pyritic Sulfur Removal from Coal of Various Ranks

	Coal		Pyritic sulfur removal % w/w	
Rank	Sample No.	Mesh size	Without bacteria	With bacteria
Bituminous	1	-100 to 200	2.0	3.0
		-200 to 325	1.9	3.7
		-325	3.3	63.2
	2	-100 to 200	3.9	14.7
		-200 to 325	2.3	53.6
		-325	3.5	76.3
	3	-100 to 200	3.0	23.9
		-200 to 325	2.5	54.4
		-325	2.8	70.8
Subbituminous	4	-100 to 200	2.8	0
		-200 to 325	2.3	0
		-325	3.6	0
	5	-100 to 200	0.7	0
		-200 to 325	0.8	0.2
		-325	2.3	10.1
Lignite	13	-100 to 200	5.8	0
		-200 to 325	0	0.8
		-325	7.5	3.0

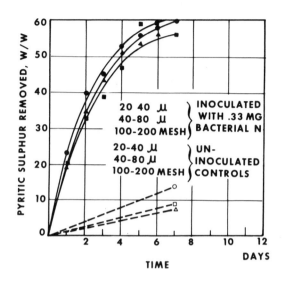

Figure 6-9. Rate of Removal of Pyritic Sulfur Kentucky No. 11
Coal in the Presence of Ferrobacillus
Ferrooxidans (13)

that subbituminous and lignite coals are essentially unaffected by this system, while bituminous coals reacts up to 70.8% removal. The authors attribute the low reactivity of subbituminous and lignite coals to the presence of acid neutralizing ash components.

REFERENCES

1.   H.H. Lowry, ed., Chemistry of Coal Utilization – Supplementary Volume, John Wiley, New York, 1963, pp 272-281.

2.   A.R. Powell and S.W. Parr, A Study of the Forms in Which Sulfur Occurs in Coal, University of Illinois Bulletin, 16, No. 34, (1919), pp 36-38.

3.   S.H. Li and S.W. Parr, Ind. and Eng. Chem., 18: (12) 1299 (1926).

4.   H.W. Nelson, R. D. Snow and D.B. Keyes, Ind. and Eng. Chem., 25: (12), 1355 (1933).

190

OXYGEN IN AQUEOUS SOLUTION

5. R.A. Meyers and J.W. Hamersma, unpublished results (1974).

6. S.S. Sareen, R.A. Giberti, P.F. Irminger and L.J. Petrovic, "The Use of Oxygen/Water for Removal of Sulfur from Coals," 80th National Meeting of the Am. Inst. of Chem. Eng., Boston, Massachusetts, 1975.

7. M. Ichikuni, Bull. Chem. Soc. (Japan), 33: 1052 (1960).

8. E.P. Koutsoukos, R. Orsini, G.J. Ogle, and R.A. Meyers, "Chemical Desulfurization of Coal by the Meyers Process," 73rd National Meeting of the Am. Inst. of Chem. Eng., Salt Lake City, Utah, August 1974.

9. D.R. McKay and J. Halpern, J. Trans. Met. Soc. of AIME, 212: 301 (1958).

10. I.H. Warren, Austral. Journ. Appl. Sci., 7: 346 (1956).

11. S. Friedman, R.B. Lacount and R.P. Warzinski, to be published in Am. Chem. Soc. Dv. Fuel Chem., 21 (1977).

12. J.C. Agarwal, R.A. Giberti, P.F. Irminger, L.F. Petrovic and S.S. Sareen, "Chemical Desulfurization of Coal," Mining Congress Jr., 61: (3), 40 (1975).

13. Z.M. Zarubina, N.N. Lyalikova, Ye. I. Shmuk, Izevest. Akad. Navk S.S.S.R., Otdel. tekh. Nauk Metal. i Topl., No. 1, 117 (1959).

14. M.P. Silverman, M.H. Rogoff and I. Wender, Fuel, 42: 113 (1963).

CHAPTER 7

PYRITIC SULFUR REMOVAL PROCESSES —
GAS-SOLID METHODS

I. Experimental Data

II. Engineering Design and Cost Estimation

The roasting or air oxidation of pyrite is shown in Chapter 3 to proceed at a measurable rate beginning about $300^{\circ}C$ and to attain near completion in 150 minutes at $400^{\circ}C$, with abundant air admission and a large specific pyrite surface (1). The product is almost entirely ferric oxide and sulfur dioxide. Muntean (2) studied the kinetics of the reaction of coal with air at $380^{\circ}C$, concluding that the rate of oxidation of pyrite in coals is approximately the same as that of mineral pyrite. The reduction of pyrite with hydrogen was shown (Chapter 3-I.C.) to proceed stepwise to form ferrous sulfide and hydrogen sulfide, beginning at $300^{\circ}C$, then iron and hydrogen sulfide near $400^{\circ}C$. The virtually complete reduction of pyrite in coal to ferrous sulfide and hydrogen sulfide was shown to occur during coal hydrogenation in anthracene oil at $400^{\circ}C$ (3).

The following two sections discuss experimental data obtained for desulfurization of coal by gas-solid processes and an engineering design and cost estimation for air oxidation.

I. EXPERIMENTAL DATA

Blum and Cindea (4) found that coal can be desulfurized in a steam/air mixture (1 hr at $380^{\circ}C$) with a carbon loss of only 4-5%. Sinha and Walker (5) performed a detailed study of the roasting of seven U.S. coals in air under essentially anhydrous conditions. Low and medium volatile coals (the Lower Kittanning and Freeport seams of the Appalachian Basin) were treated as well as high volatile coals from the Eastern Interior Basin, such as the Illinois No. 6 seam. Experimentally, a current of dry air was passed over a coal sample contained in a silica boat in the center of a tube furnace. The extent of sulfur removal was followed by either measurement of the total sulfur content of the coal before and after treatment or trapping of the sulfur dioxide product. Sinha and Walker believed that removal of organic sulfur was small, since it was not accessible to oxygen except where gasification of the carbon skeleton in the coal occurred.

191

Initial studies demonstrated that treatment of 100 mesh x 0 coal resulted in higher removals of pyritic sulfur than treatment of 40 mesh x 70 or 70 mesh x 0 size fractions. Roasting in air for 15 minutes at $400^{\circ}C$ resulted in sulfur reductions corresponding to 16-42% removal of the pyritic sulfur (Table 7-1), assuming no organic sulfur reduction. Coal volatilization losses of 1-13% and residual coal heat content losses of 12-20% were found (both of the latter increasing with decreasing rank). Thus, a part of the coal was burned to CO and CO_2, while a considerable portion of the organic coal matrix was partially combusted during the roasting process through formation of oxygenated functional groups still attached to the coal matrix.

TABLE 7-1

Air Oxidation: Pyritic Sulfur Removal and Product Properties*(5)

Sample No.	Seam/state	Pyritic sulfur removed, % w/w	Coal Wt loss % w/w	Residue heat content loss % w/w
1	Lower Kittanning/Pa	16	1.0	11.6
2	Freeport/Pa	49.6	4.0	14.3
3	Illinois No. 6/Ind	35.0	8.5	17.7
4	Illinois No. 6/Ill	42.0	13.0	19.5

*100 mesh x 0, 15 mins @ $400^{\circ}C$.

The effect of reaction time on four samples is shown in Table 7-2; the data indicates that one hour of treatment increases pyritic sulfur removal. Maximum removal was obtained at $450^{\circ}C/10$ minutes (Table 7-3) with only slightly higher coal weight losses. Not explained was an apparent reduction in coal heat content loss at the higher temperature (with the exception of Sample 4, in which there was a higher heat content loss).

TABLE 7-2

Air Oxidation: Pyritic Sulfur Removal *vs* Reaction Time*(5)

Sample No.	Native pyritic sulfur % w/w	Pyritic sulfur removed, % w/w		
		20 min	30 min	60 min
1	2.0	40.0	64.0	67.5
2	2.6	71.6	77.6	87.6
3	1.6	62.0	−	−
4	4.2	49.0	−	−

*100 mesh x 0, $400^{\circ}C$

TABLE 7-3 (5)

Air Oxidation: Maximum Pyritic Sulfur Removal*

Sample No.	Pyritic sulfur removed % w/w	Wt loss % w/w	Heat content Btu/lb		lbs S/10^6 Btu		Selectivity**
			Native	Treated	Native	Treated	
1	88.2	4.5	13745	12580	2.1	0.9	0.4
2	77.0	5.5	13233	12064	3.1	1.8	0.4
3	108.0	13.5	12126	10242	2.7	1.7	0.4
4	52.4	16.7	12773	9250	5.4	5.4	0

*100 mesh x 0, 10 mins @ 450°C

**$\dfrac{N - T}{N}$

These results indicate that roasting of dry coal in air for 10 minutes at 450°C can give a coal product in 85-95% yield, from which 50-100% of the pyritic sulfur is removed. However, the product heat content is reduced by about 10-25% over the native coal. In no case was the total sulfur content reduced below 1.2%; when sulfur reductions are considered on the basis of sulfur content per unit heat content (Table 7-3) it can be seen that, for an Illinois No. 6 coal (Sample 4), a 52.4% reduction in pyritic sulfur in the product results in no decrease in the sulfur content per unit heat content because of a steep reduction in the heat content of the product. A calculated value of a selectivity function for sulfur removal per unit heat content from Sinha and Walker's data is 0.4 for the first three coals (Table 7-3) and is nil for the last coal (Sample 4).

Overall, the best results were obtained with Appalachian coals (Lower Kittanning and Freeport) in which nearly maximum pyritic sulfur removal was obtained, while weight losses were held to nearly 5% and heat content reductions to about 10%.

Jacobs and Mirkus (6) studied the desulfurization of an Illinois No. 6 coal with stream-air mixtures in a fluidized bed reactor. They found that 20-60 minutes at 510°C gave maximum sulfur removal (60-70%) and that pyritic sulfur rather than organic sulfur was decomposed. Uncorrected sulfur analysis of the coal residue tended to indicate higher removal than actual, since coal gasification left a residue diluted by ash. Sulfur removal corrected for ash dilution of the residue (Figure 7-1) shows that sulfur partition between the residue and gas phases begins at about 200°C and reaches a sulfur gasification maximum between 400 and 600°C. No heat content data was given.

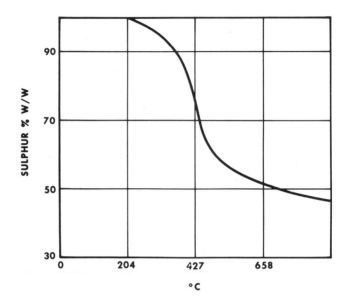

Figure 7-1. Fluidized Bed Air Treatment: Corrected Sulfur
Content of Residue *vs* Treatment Temperature

Block, Sharp and Darlage (7) studied the effect of air, steam and hydrogen on the desulfurization of 10 U.S. high-volatile bituminous coals. Their experimental method, similar to that of Sinha and Walker, involved the passage of air over a boat containing coal in an electric tube furnace. Organic sulfur in the coal was determined by "Leco" analysis for total sulfur of a sample previously extracted with nitric acid to remove pyrite. The difference between the coal sulfur analysis and the organic sulfur as determined above was designated inorganic (pyritic) sulfur. All coal samples were initially pretreated to destroy their swelling properties by heating in an electric oven at $300^{\circ}C$ for ten minutes in the presence of air.

Treatment with air at $450^{\circ}C$ for 15 minutes resulted mainly in the attack of the pyritic sulfur (a discussion of the organic sulfur removal results is presented in Chapter 10). Contrary to the results of Sinha and Walker, the pyritic sulfur was only incompletely removed (Table 7-4). Further, weight losses were high and the fuel value of the residue was significantly reduced. The authors concluded that treatment at $450^{\circ}C$ is not practical.

TABLE 7-4

Air Oxidation: Maximum Sulfur Removal* *vs* Product Properties

Sample No.	Coal Seam or Identification	Pyritic sulfur change % w/w	Wt loss % w/w	Residue heat content loss % w/w	Total sulfur % w/w		lbs S/10^6 Btu		Selectivity**
					Native	Treated	Native	Treated	
1	Pittsburgh No. 8 (Belmont Co.)	-59	34	15	3.59	2.37	2.8	2.2	0.1
2	Pittsburgh No. 8 (Jefferson Co.)	-80	15	5	1.79	0.91	1.3	0.7	0.5
3	Illinois No. 6	-19	19	7	4.08	3.27	3.9	3.4	0.1
4	Illinois No. 5	-16	19	6	1.82	1.09	1.3	0.9	0.3
5	Howard Co./Missouri	-68	18.1	16	3.90	2.02	3.1	1.9	0.4
6	MacFarlane Mine/Utah	+56	42	25	5.23	4.51	5.5	6.4	-0.2

*10 mins @ 450°C, 325 mesh x 0.

**$\dfrac{N - T}{N}$

Indeed, if one calculates a selectivity function (Table 7-4), it can be seen that while the values which were calculated from the data found by Sinha and Walker were approximately 0.4 (Table 7-3), the selectivity for sulfur removal per unit heat content for the method of Block et al tends to be somewhat lower, although not in all cases. One Pittsburgh No. 8 seam coal (Sample No. 2) showed weight loss, heat content loss and selectivity similar to that found by Sinha and Walker. Moreover the sulfur content per unit heat content was 0.7, which is close to the 0.6 required to meet the Federal Standards of New Stationary Sources (8). Thus, for this single Pittsburgh No. 8 coal it would appear that a viable process existed. (See the next section.)

Steam treatment was most effective at $600^{\circ}C$ for an average removal of 61% of the total sulfur and 87% of the pyritic sulfur. Calculations (Table 7-5) show that two of the samples (Nos. 2 and 4) met the New Stationary Source Standards after treatment, and in general that the selectivity for steam treatment was higher than the selectivity found for air treatment. Weight losses were on the order of 25-30%. Thus, steam treatment at $600^{\circ}C$ could conceivably be contemplated as a method for desulfurization of coal. The major engineering drawbacks to a process based on steam treatment, however, lay in the nature and value of the volatilized products and in the process heat balance. Although air treatment could well provide its own process heat, the steam treatment would require external heating. A combination of air and steam could possibly provide a worthwhile engineering process provided the selectivity could be maintained.

Hydrogen was not effective below $850^{\circ}C$, but at $900^{\circ}C$ (Table 7-6) 80-90% of the total sulfur was removed, but with weight losses of 20-50%. The initial oxidative pretreatment was quite effective in lowering required hydrogenation reaction time. Even though the matrix must have been heavily hydrogenated, the heat content of the treated residues tended to remain the same or decrease. Evidently the oxidative pretreatment created reducible sites in the coal matrix which tended to restrict heat content built up. Sulfur contents were reduced to levels below that required to meet the Federal standards at the product heat content. The development of a process based on this type of gas-solid hydrogenation is thus limited by the extreme weight losses and probable high consumption of hydrogen.

The treatment of coal with carbon monoxide and $CO-H_2O$ gas mixtures at $400-600^{\circ}C$ was investigated by Sinha and Walker (9). They found reductions in total sulfur content after treatment, but in no case was the char sulfur content less than 1.3% w/w. It was concluded that the sulfur removal attained was not sufficiently large to be of commercial interest.

TABLE 7-5

Steam Treatment: Maximum Sulfur Removal* vs Product Properties

| Sample No. | lbs S/10^6 Btu | | Selectivity** |
	Native	Treated	
1	2.8	1.5	0.5
2	1.3	0.6	0.5
3	3.9	1.2	0.7
4	1.3	0.6	0.5
5	3.1	1.1	0.7
6	5.5	4.1	0.3

*15 mins @ 600° C, 325 mesh x 0.

**$\dfrac{N - T}{N}$

TABLE 7-6

Hydrogenation: Maximum Sulfur Removal* vs Product Properties

| Sample No. | Sulfur content, % w/w | | Wt loss % w/w |
	Native	Treated	
1	3.59	0.36	38
2	1.79	0.74	21
3	4.08	0.38	36
4	1.82	0.35	38
5	3.90	0.45	52
6	5.23	0.57	44

*4 mins @ 900°C after 10 mins @ 300°C in air.

II. ENGINEERING DESIGN AND COST ESTIMATION

Battelle researchers (10) considered the engineering design of a process based on the preferential air oxidation of coal (Figure 7-2). The following conditions were assumed in the design: 1) a fluidized bed operating at 350-460°C is used; 2) coal, with an input heat content of 12,500 Btu/lb is fed at a rate of 384 tons/hr (high sulfur, Eastern coal); 3) only pyritic sulfur is removed; 4) residence time in the fluidized bed is one hour; 5) air flow rate is twice stoichiometric for conversion of pyritic sulfur to sulfur dioxide and iron oxide, and that this is well above the requirement for minimum fluidization velocity; 6) coal particle size is 150 mesh x 0; 7) the process is associated with a pulverized coal fired boiler.

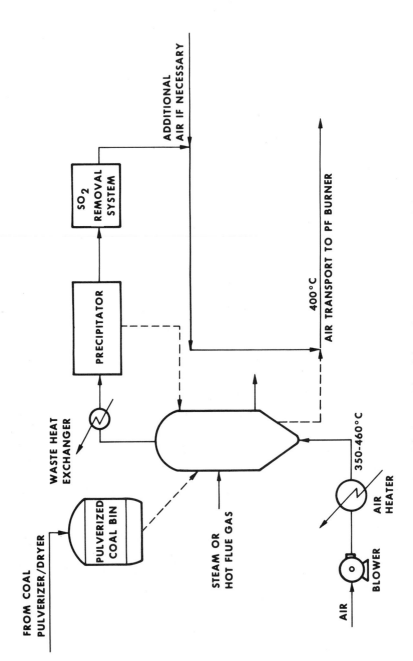

Figure 7-2. Process Flow Scheme for Preferential Air Oxidation
of Sulfur in Coal

Although no experimental basis was given for the above assumptions, it is likely that the temperature selection was based on the work of Sinha and Walker (5) or Block, et al (7) and not on the fluidized bed research of Jacobs and Mirkus (6) who used higher temperatures. The one-hour residence time is puzzling in view of the data obtained utilizing 10 minutes at 450°C. It is possible that this residence time was based on the 20-60 minutes at 510°C advocated by Jacobs and Mirkus. No mention is made in the design of volatile products which would be expected as a result of the oxidative treatment for an Eastern coal (see Samples Nos. 1 and 2, Table 7-3 and Samples No. 1 and 2, Table 7-4), nor does the equipment list (Table 7-7) include provision for condensation of volatiles.

TABLE 7-7

Air Oxidation Costs (10)

Equipment	Capacity	Installed Cost ($)
Blower	21,000 fm, 100 hp	250,000
Air Heater	130 ft^2	10,000
Fluidized Bed and Heat Exchanger	21 ft diameter by 63 ft high 10,000 ft^2 heat exchange area	2,200,000
Waste Heat Exchanger	130 ft^2	10,000
Precipitator	27,000 fm	60,000
SO$_2$ Removal System	280 tons SO$_2$/day	6,000,000
Subtotal		8,530,000
Contingencies @ 15% of installed cost		1,280,000
Contractor fee @ 3% of installed cost		260,000
		10,070,000

	Operating Costs
Item	$/yr
Operating Labor	144,000
Supervision	52,000
Maintenance	500,000
Overhead	120,000
Steam	600,000
Electricity	5,000
Raw Materials	650,000
Taxes and Insurance	200,000
	2,271,000

The Battelle researchers charge the cost of coal grinding to the associated power plant and derive process heat by exchange with steam and/or flue gas generated in the power plant. Capital and operating cost estimates (Table 7-7) are $10 x 10^6 and about $0.30/10^6 Btu, respectively.

REFERENCES

1. G.M. Schwab and J. Philinis, J. Am. Chem. Soc., 69, 2588 (1947).

2. V.C. Muntean, Akad. Rep. Populare Romine Studii Cercetari Met., 8, 331 (1963).

3. R.A. Meyers, J.W. Hamersma, R.M. Baldwin, J.G. Hardwerk, J.H. Gary, J.O. Golden, Am. Chem. Soc. Dv. Fuel Chem. Preprints, 20 (1), 234 (1975).

4. I. Blum and V. Cindea, V. Pop. Romine Inst. Energ. Studii, 11, 325 (1961).

5. R.K. Sinha and P.L. Walker, Jr., Fuel, 51, 126 (1972).

6. T.K. Jacobs and J.D. Mirkus, Ind. and Eng. Chem., 50: 24 (1958).

7. S.S. Block, J.B. Sharp and L.J. Darlage, Fuel, 54: 113 (1975).

8. U.S. Environmental Protection Agency, Federal Register, 36: 24876 (1973).

9. R.K. Sinha and P.L. Walker, Jr., Fuel, 51: 329 (1972).

10. Liquefaction and Chemical Refining of Coal, A Battelle Energy Program Report, Columbus, Ohio, July (1974).

CHAPTER 8

PYRITIC SULFUR REMOVAL PROCESSES —
CAUSTIC LEACHING

I. Molten Caustic

II. Aqueous Caustic

The reaction of mineral pyrite with caustic to form sodium sulfide, sodium thiosulfate and iron oxide (Eq. 1) was discussed in Chapter 3. Since the products are soluble in the caustic extraction solution, pyrite can be removed from coal using this method. The extraction of pyrite from coal with caustic can be performed using either molten caustic salts or aqueous solutions; these two methods are discussed in detail in the following sections. Organic sulfur removal results for the reaction of caustic with coal are discussed in Chapter 10.

$$8\ Fe\ S_2 + 30\ Na\ OH \rightarrow 4\ Fe_2\ O_3 + 14\ Na_2\ S + Na_2\ S_2\ O_3 + 15\ H_2O \quad (1)$$

I. MOLTEN CAUSTIC

Masciantonio (1,2) studied the effect of molten caustic on pyrite in bituminous coal. He found that rapid reaction occurs between coal pyrite and caustic at temperatures near $250^{\circ}C$, and that the pyritic sulfur is converted to sulfides which are soluble in the molten caustic. The caustic may be filtered from the coal and the coal treated with water to remove residual salts. Caustic salts which are suitable include anhydrous sodium hydroxide, potassium hydroxide, sodium or potassium acetate or mixtures of these. Lime melts at a temperature which is too high for compatibility with the coal, but can be used as a diluent. Most of the actual work was performed with a 1:1 melt of sodium hydroxide and potassium hydroxide because of its thermal stability and relatively low melting point (about $300^{\circ}C$).

Masciantonio found that at temperatures between $150\text{-}225^{\circ}C$ only pyritic sulfur is removed from the coal, and below $150^{\circ}C$ no measurable removal of either type of sulfur occurs. Sulfur removal was based on the differential sulfur content between treated coal and residue. As indicated in Table 8-1 sulfur removal is at a maximum at approximately $400^{\circ}C$ for the Pittsburgh seam coal investigated, but there is a significant decrease in coal volatile matter and loss of coal-swelling properties. Similar results were obtained with a Wyoming and an Illinois coal, but with larger losses of coal due to hydrolysis. Coal particle size up to 1/4 in. x 0 was not an important factor since the coal becomes "plastic" at about 325°C under these conditions. Pittsburgh seam coal gave a yield of 89-93%, while the Illinois and Western coal gave only 69-76% and 52% yield, respectively. The mechanism of the caustic hydrolysis of coal, as well as independent measurements of the degree of coal hydrolysis in caustic, are presented in Chapter 10.

TABLE 8-1

Molten Caustic Treatment of Coal*: Comparison of Properties

Treatment	Sulfur % w/w	vm % w/w	fsi
Native Coal	1.6	36.1	6-1/2
30 mins @ 150°C	1.6	35.2	5-1/2
30 mins @ 300°C	1.0	33.7	3-1/2
30 mins @ 350°C	0.8	32.2	3
30 mins @ 400°C	0.5	23.6	0

*Pittsburgh seam

Ultimate analysis of the Pittsburgh seam coal after treatment showed small carbon and hydrogen losses but an increase in the sum of nitrogen and oxygen content (Table 8-2). Unfortunately, data on the heat content of treated residue was not given so that an evaluation of the selectivity of the process for sulfur content removal per unit heat content cannot be made.

TABLE 8-2

Molten Caustic Treatment of Coal*: Ultimate Analyses (2)

Treatment	Carbon % w/w	Hydrogen % w/w	Sulfur % w/w	Oxygen and nitrogen % w/w	Atomic H/C
Native Coal	84.7	5.7	1.8	7.9	0.80
30 mins @ 200°C	85.3	5.7	1.2	7.8	0.80
30 mins @ 350°C	83.8	5.7	0.9	9.7	0.81
30 mins @ 450°C	82.8	3.7	0.6	12.9	0.53

*Pittsburgh seam

A process schematic is shown in Figure 8-1 (2). Suitably ground coal is fed to a reactor along with a fused salt (residence time, up to 45 mins) where pyrite and some of the organic sulfur is claimed to form sodium and potassium sulfides which dissolve into the fused salt. (No mention is made by Masciantonio of the expected (Eq. 1) thiosulfate and iron oxide salts.) The slurry is separated from the coal by allowing the molten caustic to settle. The fused salt thus separated (along with increasing amounts of iron, thiosulfate salts, as well as probably other ash components) is recycled, while the desulfurized coal containing sorbed (and possibly chemically-bonded) alkali salts, is contacted with fresh wash water. This slurry is filtered and the wash water is evaporated to provide additional fused salt and purified water for recycle. The desulfurized and washed coal is fed to a storage system.

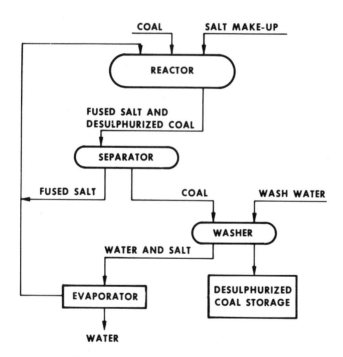

Figure 8-1. Molten Caustic Flow Scheme (2)

Masciantonio also did not contend with the hydrolysis products which form on treatment of coal with caustic. The nature of these hydrolysis products, which are phenols and acids, neutral oils, hydrocarbon gases and carbon dioxide, are discussed in detail in Chapter 10. It is noteworthy that Pittsburgh seam coal (3), treated with 100% sodium hydroxide at 350°C for 20-30 hrs, gives a 1.1% yield of carbon dioxide and 0.7% yield of phenols and acids, based on carbon content. This degree of hydrolysis, which would be an upper limit for the work reported here, is quite small compared to that obtained for aqueous caustic solutions. Carbonation of a portion of the aqueous alkali solution when leaving the wash (Figure 8-1), followed by extraction of the neutralized phenols and acids with an organic solvent, would allow removal of the hydrolysis products. Carbon dioxide could be vented and cleaned.

II. AQUEOUS CAUSTIC

Reggel, Raymond, Wender, and Blaustein (4) have reported the desulfurization of Eastern Interior Basin coal by treatment with 10% aqueous sodium hydroxide. They found that pyritic sulfur is largely removed, but the organic sulfur in the coal is not attacked by their procedure and in some cases an erratic increase in organic sulfur content was found (see Chapter 10). A coal yield of 91.5% (maf basis) was obtained for the Illinois No. 6 seam coals investigated. Approximately 45 to 95% of the pyritic sulfur was removed (Table 8-3) as determined by differential pyritic sulfur analysis of the native and treated coals after acidification with carbon dioxide, hydrochloric acid, sulfur dioxide or sulfuric acid. A large fraction of the ash was removed during the acidification process, most likely due to neutralization of alkali salts of alumino-silicates which had been formed as a result of the sodium hydroxide treatment. For example, an Illinois No. 6 coal containing 9.8% ash was reduced to a product containing 0.7% ash after acidification. Increasing the extraction temperature from the standard 225° to 325°C did not dramatically increase the amount of pyritic sulfur removal.

The reported procedure for the Battelle hydrothermal process (5) appears similar to the treatment reported by Reggel et al (4) with the exception of the addition of a small amount of calcium hydroxide to the leach solution. The Battelle process utilizes a 10% w/w sodium hydroxide/2% w/w calcium hydroxide solution to treat pulverized coal at temperatures between 225 and 350°C (pressures of 350 to 12,500 psi) for periods up to 30 minutes. This would compare to the standard conditions utilized by Reggel et al of 10% w/w sodium hydroxide treatment at 225 to 325°C for two hours. Thus, the major difference in experimental conditions appears to be a shorter residence time and the addition of calcium hydroxide.

TABLE 8-3

Inorganic Sulfur Removal on Treatment of Coal with
Aqueous Sodium Hydroxide* (4)

Mine**	Treatment	Workup	Sulfur content, % w/w			Pyritic sulfur removal %
			Total	Sulfate	Pyrite	
River King	Native Coal	—	3.33***	0.29	1.05	—
	NaOH	CO_2	2.05	0.11	0.13	88
	$Ca(OH)_2$	HCl	3.04	0.04	1.04	1
	NaOH	HCl	2.54†	0.01	0.11	90
	H_2O	CO_2	2.77	0.01	0.98	7
	H_2O	HCl	2.85	0.01	1.04	1
	NaOH	SO_2	2.40	0.23	0.19	82
	NaOH	H_2SO_4	2.75	0.24	0.19	82
Elliot	Native Coal	—	4.76††	0.28††	3.53††	—
	NaOH	CO_2	2.27	0.19	0.19	95
	NaOH	HCl	3.61†	0.01	1.94	45
	NaOH, 325°	HCl	2.77	0.04	0.48	86
	NaOH	H_2SO_4	5.21†	0.06	3.35	5

*10% aqueous NaOH for 2 hrs at 225°C followed by acidification (workup).

**200 mesh x 0.

***Average of six samples.

†Organic sulfur increased.

††Average of four samples.

Experimental results (6) for the extraction of pyrite from four coals are shown in Table 8-4. Pyritic sulfur removal was determined on the basis of the differential pyritic sulfur content of the native and treated coal (maf basis). The Battelle researchers contended that organic sulfur is also removed during this process, which seems contradictory to the erratic organic sulfur content increase reported by Reggel et al.

Acidification of the aqueous solution is not standard practice; rather, the caustic is centrifuged away from the coal and the coal is washed with water. An optional treatment of product coal with dilute acid would extract additional ash. Overall sulfur reduction results (Table 8-5) indicate that the sulfur content of the coal residue remaining after caustic extraction meets the New Stationary Source Standards of 1.2 lb $SO_2/10^6$ Btu.

TABLE 8-4

Extraction of Pyritic Sulfur with Aqueous Sodium and Calcium Hydroxide (6)

			Pyritic sulfur content** %		Pyritic sulfur removal %
Experiment	Coal Mine*	Seam	Native	Treated	
1	Belmont	No. 6	1.6	0.1	92
2	Ken	No. 14	2.1	0.2	92
3	Beach Bottom	No. 8	1.7	0.1	95
4	Eagle 1	No. 5	1.5	0.2	87

*70%, unus 200 mesh.

**maf basis.

TABLE 8-5

Extraction of Sulfur with Aqueous Sodium and Calcium Hydroxide

			Sulfur content % w/w		SO_2 equivalent lbs/10^6 Btu	
Experiment	Coal Mine*	Seam	Native	Treated	Native	Treated
1	Westland	Pittsburgh No. 8	1.8	0.8	2.6	1.2
2	Beach Bottom	Pittsburgh No. 8	2.7	0.6	4.6	0.9
3	Montour No. 4	Pittsburgh	2.3	0.4	3.4	0.7
4	Martinka No. 1	Lower Kittanning	1.1	0.4	2.2	0.9
5	Martinka	Lower Kittanning	2.8	0.8	4.0	1.1
6	Renton	Upper Freeport	1.3	0.5	2.4	0.9
7	Renton	Upper Freeport	1.2	0.6	2.4	0.8
8	Sunny Hill	Ohio No. 6	2.3	0.8	3.9	1.2

*70% minus 200 mesh.

Neither specific experimental conditions nor heat content for the tabulated experiments are given. However, the heat content of both the native and treated coals can be calculated from the data given in the last two columns of Table 8-5. When this is done (Table 8-6), it can be seen that the heat content of the treated coal appears to fluctuate between a 16% decrease and a 50% increase, and for two separate treatments of coal from the same mine (Renton), one a laboratory experiment and the other a pilot plant treatment, a variation is seen from essentially no heat content change to an astounding 50% increase.

TABLE 8-6

Extraction of Coal with Aqueous Sodium and Calcium Hydroxide:
Calculated Heat Content Effects

Experiment	Coal Mine	Calculated* heat content lbs/10^6 Btu		Heat content change % w/w
		Native	Treated	
1	Westland	13,900	13,300	- 4
2	Beach Bottom	11,800	13,300	+13
3	Montour	13,500	11,400	- 16
4	Martinka No. 1	10,000	8,900	- 11
5	Martinka	14,000	14,500	+ 4
6	Renton	10,800	11,100	+ 3
7	Renton	10,000	15,000	+50

$$*\text{Coal heat content} = 2 \times \frac{\% \text{ sulfur in coal}/100}{\text{lbs } SO_2/10^6 \text{ Btu}} \times 10^6.$$

As pointed out previously, Kasehagen (3) studied the action of aqueous sodium hydroxide on a Pittsburgh seam coal at temperatures of 250-400°C, various alkali concentrations and reaction times of 20-30 hrs. He reports the decomposition of the Pittsburgh seam coal into a coke-like residue and phenols, acids, neutral oils, hydrocarbon gases and carbon dioxide. Clearly, the extent of decomposition reported in his results forms an upper limit for the type of decomposition which might be expected during the Battelle Hydrothermal treatment. For example, Kasehagen reports a yield (based on carbon balance) of approximately 9% phenols and acids, 4% carbon dioxide, nearly 2% hydrocarbon gases and 1% neutral oils from treatment of the Pittsburgh seam with 10% sodium hydroxide at 350°C for 20-30 hrs. Thus, an 84% yield of coal carbon content would be an approximate minimum in the Battelle process, while a 95% yield is reported. Kasehagen released the alkali soluble phenols and acids by treatment with carbon dioxide, a second step identical to that used by both Reggel *et al* and Battelle.

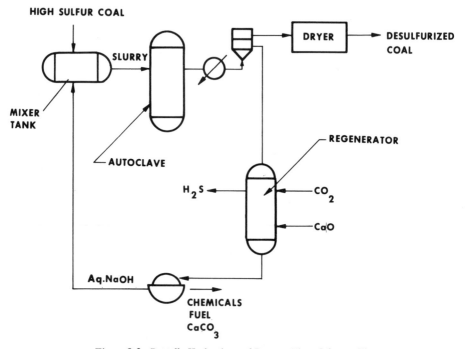

Figure 8-2. Battelle Hydrothermal Process Flow Scheme (5)

A process schematic is shown in Figure 8-2 for the Battelle Hydrothermal process (5). Suitably prepared and crushed coal (70% -200 mesh) is mixed with caustic leaching solution in a slurry tank. After mixing, the slurry is heated under pressure to reaction temperature (in apparently a batch mode). The slurry is centrifuged to remove the leachant, and either given an optional post-treatment washing and deashing with acid or dried directly for use in a boiler. Apparently, residual caustic would be left on the coal filter cake where no post-treatment was used, adding to the makeup burden of the process. The leachant solution is regenerated by first, carbonation to convert the sodium sulfide, formed from pyrite, to hydrogen sulfide and sodium carbonate (and at the same time neutralize the remaining sodium hydroxide and calcium hydroxide). In accordance with Kasehagen (3), the extracted phenols and acids would be neutralized at this point and would probably form a separate layer which could be removed from the aqueous solution by skimming or solvent extraction along with undissolved oils. The resulting hydrogen sulfide is handled in a Claus Unit where it is oxidized to elemental sulfur and water. Regeneration of sodium hydroxide is accomplished by addition of lime, which converts the sodium carbonate to sodium hydroxide and calcium carbonate. The sodium hydroxide is recycled and lime can be regenerated by thermal treatment to form carbon dioxide. Chemicals, fuels and metals are shown leaving the process during the filtration of the regenerated leach solution. No mention is made of thiosulfate salts (Eq. 1) which must be removed from the leach liquor, if indeed formed.

Battelle estimated a cost of $10-15 a ton for the process in a full size plant. Information is not given as to the method of arriving at this cost or whether the cost applies to input coal or output coal.

REFERENCES

1. P.X. Masciantonio, Fuel, 44: 269 (1965).

2. P.X. Masciantonio, U.S. Patent 3,166,483 (1965).

3. L. Kasehagen, Ind. Eng. Chem., 29: 600 (1937).

4. L. Reggel, R. Raymond, I. Wender, B.D. Blaustein, Am. Chem. Soc., Fuel Dv. Preprints, 17: (1), 44 (1972).

5. Science, 189: 129 (1975).

6. E.P. Stambaugh, J.F. Miller, S.S. Tam, S.P. Chauhan, H.F. Feldman, H.F. Carleton, J.F. Foster, H. Nack, and J.H. Oxley, Hydrocarbon Processing, 115 (July 1975).

CHAPTER 9

PYRITIC SULFUR REMOVAL PROCESSES — MISCELLANEOUS OXIDANTS AND REDUCTANTS

I.   Sulfur Dioxide

II. Nitric Acid

III. Hydrogen Peroxide

IV. Chlorine

V.   Potassium Nitrate

VI. Reduction

The use of ferric sulfate, oxygen, hydrogen and caustic has received considerable attention in studies for the removal of pyritic sulfur from coal. Five additional chemical treatments have been evaluated in some preliminary studies: sulfur dioxide, nitric acid, hydrogen peroxide, potassium nitrate and chlorine. In addition, some aqueous reducing agent systems are proposed. These are discussed in the six sections to follow.

I. SULFUR DIOXIDE

Sulfur dioxide has sufficient electrochemical potential to oxidize pyrite under standard conditions (Chapter 3-I.B.). Aqueous sulfur dioxide in strong hydrochloric acid solution has been shown to oxidize 200 mesh x 0 iron pyrite (1) in 55% yield in a period of 24 hours at $180^{\circ}C$ (2). The reaction products are ferrous chloride and elemental sulfur (Eq. 1). This reaction would be useful for removal of pyritic sulfur from coal provided the elemental sulfur generated by the reaction could be removed. It has been shown in Chapter 5 that elemental sulfur can be removed from coal by vaporization, solvent extraction or possibly by chemical reaction. There remains the necessity to regenerate both the sulfur dioxide and hydrochloric acid and to recover iron dissolved in the leach solution. Ferrous chloride can be oxidized to ferric oxide and hydrochloric acid (Eq. 2) during pyrolytic air oxidation (3,4). Thus, the iron obtained from pyrite could be reclaimed as iron oxide, and the hydrochloric acid recycled. Sulfur is easily oxidized to sulfur dioxide with oxygen (Eq. 3), so that the net reaction could be written as the reaction of oxygen with pyrite to give iron oxide and elemental sulfur (Eq. 4).

$$FeS_2 + \frac{1}{2}SO_2 + 2\ HCl \rightarrow Fe\ Cl_2 + \frac{5}{2}S + H_2O \tag{1}$$

$$H_2O + Fe\ Cl_2 + \frac{1}{4}O_2 \rightarrow \frac{1}{2}Fe_2\ O_3 + 2\ HCl \tag{2}$$

$$S + O_2 \rightarrow SO_2 \tag{3}$$

$$FeS_2 + \frac{3}{4}O_2 \rightarrow \frac{1}{2}Fe_2\ O_3 + 2\ S \tag{4}$$

Some feasibility studies were performed by Meyers, Hamersma and Kraft (5) on an Indiana No. 5 coal at temperatures between 100-140°C/20-30 psig (Table 9-1). These lower temperatures were chosen to minimize chlorination by $SOCl_2$ (in equilibrium with SO_2 and HCl) of coal. Analysis of total sulfur content of the native and treated coals indicated 30-52% removal of pyritic sulfur. Substantial ash reduction occurred due to the solubilizing nature of the combination of sulfur dioxide and hydrochloric acid with an appropriate heat content increase for Experiments 1 and 3 for a 4% w/w ash decrease.

Experimentally, 0.9 M aqueous sulfurous acid (10 x stoichiometric excess over pyrite content of coal) in 3.6 M hydrochloric acid and pulverized coal was introduced into a glass aerosol bomb fitted with a magnetic stirrer and pressure gauge. The slurry was heated to temperature (at 20-30 psig) in an oil bath and maintained for 20 hours, then cooled and filtered from the aqueous phase. All samples were washed and then dried in a vacuum over to constant weight. Most of the elemental sulfur which had been formed vaporized from the coal under these conditions. The coal from Experiment No. 3 was given a post-extraction with benzene to remove any remaining elemental sulfur, resulting in a slightly higher total sulfur removal. The higher temperature run (No. 1) gave less removal than extraction at 100°C possibly due to consumption of $SOCl_2$ by the coal matrix.

Although higher temperatures and/or longer retention times could conceivably give more nearly complete pyritic sulfur removal, there may possibly be more side reaction. Pulverization to 325 mesh x 0 to give more pyrite surface area would be a better approach to increasing desulfurization.

TABLE 9-1

Sulfur Dioxide and Hydrochloric Acid Pressure Leaching of Coal*

Experiment No.	Extraction temp °C	Total sulfur content % w/w	Ash % w/w	Heat content Btu	Total sulfur reduction %	Pyritic sulfur reduction %	Ash reduction %	Btu change %
Native coal		3.62	11.8	12,500	–	–	–	–
1	140	3.06	7.4	13,000	15	30	37	+4
2	100	2.76	7.8	12,600	23	46	33	+1
3**	100	2.69	7.5	13,000	16	52	36	+4

*Leach solution 3.6M in HCl, 0.9M in SO_2, 20 hr reaction time, 100 mesh x 0 coal, washed with hot water to remove residual acid, coal dried at 160°C in vacuum oven to constant weight.

**Washed with hot benzene for 20 mins, then dried as above.

A process schematic is shown in Figure 9-1. High sulfur thoroughly pulverized coal is mixed with aqueous hydrochloric and sulfurous acids which wet the coal in a mixing tank. The slurry is transferred to a pressure reactor operating at 100-120°C. Sulfur dioxide is flashed from the slurry and recycled during pressure letdown and the slurry is filtered. The coal, containing elemental sulfur, is washed with water to remove residual ferrous chloride and hydrochloric acid and then dried to give a low sulfur coal product. The filtrate, containing ferrous chloride and hydrochloric acid, is reacted with oxygen at high temperature to give iron oxide product and aqueous hydrochloric acid which is recycled. A portion of the sulfur obtained on drying the coal is oxidized for sulfur dioxide recycle.

There are several drawbacks to a process based on this scheme: 1) the reactor, operating at relatively high temperature and pressure, would be susceptible to chloride corrosion, 2) there would be possible partial disproportionation of the SO_2 reagent to sulfur and sulfate, and 3) there is a risk of contamination of coal with chlorine which may become bonded to the coal matrix by the thionylchloride reaction. Switching from strong HCl to H_2SO_4 as a source of acid would certainly result in sulfonation of the coal matrix.

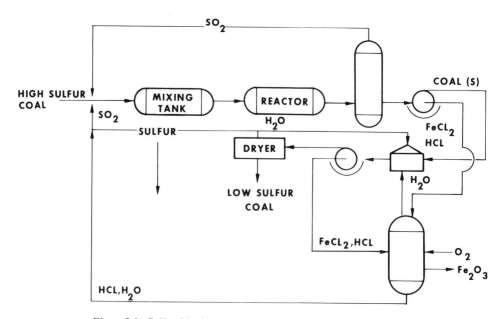

Figure 9-1. Sulfur Dioxide Pressure Leaching of Coal – Flow Schematic

Advantages of this process are: 1) conversion of reacted pyritic sulfur to the elemental sulfur form which is the most stable and desirable byproduct from a pollution control standpoint, and 2) the recyclability of the reagents.

II. NITRIC ACID

The strong oxidation potential of nitric acid relative to pyrite was discussed in Chapter 3. The extraction of pyritic sulfur from coal was studied by Powell and Parr (6) who treated Illinois coal with concentrated nitric acid, obtaining near quantitative removal of pyritic sulfur. However, they report that even at room temperature, organic sulfur compounds in coal are attacked and the organic matrix is degraded. Extraction with dilute nitric acid at higher temperatures, while effective in removal of pyritic sulfur, caused oxidation of the organic matrix. This type of extraction has since become standard for the analytical determination of coal pyrite. Mott (7) found that reaction of nitric acid with organic sulfur, and hence, the organic coal matrix, could be minimized by utilizing the following conditions: 24 hours at room temperature, aqueous nitric acid at 1.12 sp gr (22% w/w nitric acid in water).

Harms (8) proposed that a process could be developed for the removal of pyritic sulfur from coal utilizing 5-15% w/w aqueous nitric acid. At lower temperatures and lower concentrations of nitric acid, elemental sulfur was formed in preference to sulfuric acid (Eq. 5) but at higher concentrations of nitric acid and higher temperatures, elemental sulfur was converted to sulfate (Eq. 6). Room temperature extraction with 15% w/w nitric acid resulted in 32% pyrite dissolution, while at 75°C, 90% of the pyrite was dissolved. Room temperature extraction with 5% w/w nitric acid dissolved only 3% of the pyrite while 30% w/w nitric acid dissolved 82% of the pyrite. On balance, the best results were obtained with 30% w/w nitric acid at 75°C. Experimentally, a German coal (Table 9-2) was first cleaned by float-sink methods, then leached with nitric acid to remove more than 90% of the pyritic sulfur along with nearly one-third of the remaining ash. Information regarding the extent of oxidation and nitration of the organic coal matrix was not given (8).

TABLE 9-2

Removal of Pyritic Sulfur from Cleaned Coal with Nitric Acid

Coal	Coal yield % w/w	Ash content % w/w	Total sulfur % w/w	Pyritic sulfur % w/w
Raw coal	–	8.90	5.11	3.60
Cleaned	41	4.20	3.20	1.75
Leached	40	1.90	1.39	0.26

$$FeS_2 + 4\,HNO_3 \rightarrow Fe\,(NO_3)_3 + 2\,S + NO + 2\,H_2O \tag{5}$$

$$6\,FeS_2 + 30\,H\,NO_3 \rightarrow 3\,Fe_2\,(SO_4)_3 + H_2\,SO_4 + 30\,NO + 12\,H_2O \tag{6}$$

Some interesting inferences can be drawn from the work of Syunyaev et al (9) who studied the decomposition of petroleum coke in boiling 60% w/w nitric acid. The removal of organic sulfur-containing compounds was followed as a function of carbon content decrease. It was found that during the first four hours of treatment, the desulfurization rate of the coke was 1.5 times as high as the carbon burn-off rate. The treatment with nitric acid led to an increase in the content of oxygen and nitrogen in the coke, up to 3% w/w after 4 hours. These conditions are certainly more vigorous than that utilized by Harms, but are indicative of the type of reactions which are always competing during the treatment of a condensed organic matrix with nitric acid.

Diaz and Guth (10) treated coal with NO_2 gas at 93°C (synthesized in situ by reaction of NO and O_2) and then washed the coal product with water. At least 50 and up to 100% of the pyritic sulfur may have been removed in this way (Table 9-3). Diaz and Guth believed that a gas-phase oxidation of pyrite by NO_2 was operable and that the subsequent water extraction played no role in the oxidation, serving only to remove product sulfate. However, gaseous NO_2, in the presence of water contained in the coal and wash solution, forms nitric acid, which would then have been a primary reactant. The extent of pyritic sulfur removal obtained by Diaz and Guth at 93°C (Table 9-3) is similar to that observed by Harms (8) at 75°C using nitric acid.

Washing of the treated coal with 10% aqueous sodium hydroxide resulted in further sulfur content reduction, possibly including some organic sulfur. This will not be discussed here as caustic treatment of coal is discussed in detail in Chapters 8 and 10.

III. HYDROGEN PEROXIDE

Hydrogen peroxide in acid solution has a standard oxidation potential (Chapter 3) approximately 1.4v more positive than that of pyrite and about 1.6v more positive than the SO_2/S couple. Thus, H_2O_2 is a very strong oxidizing agent capable of oxidizing pyritic sulfur to the sulfate form.

Mukai et al (11) studied the treatment of several Japanese bituminous coals with 3% w/w aqueous hydrogen peroxide. This treatment was claimed to give nearly quantitative removal of pyritic sulfur without changing the caking properties of the coal.

TABLE 9-3

Coal* Desulfurization: Nitrogen(II) Oxide and Oxygen Gas —
Aqueous Extraction**

Coal	NO concentration in gas % V/V	Total sulfur % w/w	Sulfur removal % of total	% of pyritic
Lower Kittanning	9	1.58	63	76
Lower Kittanning	4.5	2.5	42	50
Lower Kittanning	5	2.9	33	39
Illinois No. 5	4.5	1.9	46	100
Illinois No. 5	10	1.9	46	100

*14 x 28 mesh coal, initial sulfur analyses: S_T, 4.3%; S_P, 3.6%; S_O, 0.7% (Lower Kittanning); and S_T, 3.5%; S_P, 1.6%; S_O, 1.9% (Illinois No. 5).

**Stationary coal bed flow-through reactor for gas treatment, 3 hrs at 93°C.

Smith (12) investigated the desulfurization of U.S. coals with hydrogen peroxide solutions. Treatment of several U.S. coals from the Appalachian and Interior Basins with 10-15% w/w H_2O_2 aqueous solution for one to two hours at ambient temperature had very little effect in removing pyritic sulfur. However, when a small amount of sulfuric acid was added to the hydrogen peroxide solutions, the reaction with pyrite was greatly enhanced. Coal (32 mesh x 0) was shaken at ambient temperature, with 5-10 v/w of aqueous H_2O_2/H_2SO_4 solution. The leach liquor was filtered from the coal and the coal was washed several times with water until all soluble sulfate was removed. The coal was then dried in a vacuum oven at 80°C to constant weight.

Treatment of a Pittsburgh seam coal with hydrogen peroxide/sulfuric acid at varying leach time (Table 9-4) showed that slightly more than 2 hours is sufficient for maximum reduction in total sulfur and ash. Sulfuric acid alone accounted for about half of the ash reduction and a small part of the total sulfur reduction. Notably the heating value of the product increased by as much as 7% indicative of little or no reaction of the hydrogen peroxide with the organic coal matrix.

As indicated in Table 9-4, sulfuric acid concentration must be maintained near 0.3 N in order to maintain high pyritic sulfur removal. Only 55% removal of pyritic sulfur is obtained utilizing 17% w/w H_2O_2 (Table 9-5) in combination with 0.1 N H_2SO_4, whereas at 0.3 N H_2SO_4 a total sulfur content reduction to 2.3% w/w is achieved (Table 9-4).

TABLE 9-4

Treatment of Pittsburgh Seam Coal with Hydrogen Peroxide –
Sulfuric Acid*

	Reagent composition					
Time	H_2O_2 % w/w	H_2SO_4 N	C	Ash % w/w	S	Heat content Btu/lb
	Native Coal		68.9	12.8	4.5	12,460
2	17	0.3	74.8	8.7	2.3	–
19	17	0.3	75.4	8.0	1.9	13,310
72	17	0.3	74.8	7.9	2.0	13,380
72	0	0.5	72.3	10.8	3.9	12,890

*100-ml solution/10g of 32 mesh x 0 coal, ambient temperature.

TABLE 9-5

Treatment of Pittsburgh Seam Coal with Hydrogen Peroxide –
Sulfuric Acid*

Reagent composition		% w/w			Sulfur forms, % w/w			
H_2O_2 % w/w	H_2SO_4 N	C	Ash	S	SO_4	Pyr	Org	Heat content Btu/lb
Native coal		68.9	12.8	4.5	0.41	2.30	1.79	12,460
7	0.1	73.0	9.5	3.3	0.06	1.44	1.80	13,150
11	0.1	72.9	9.3	3.1	0.01	1.31	1.78	13,130
15	0.1	73.3	9.3	2.8	0.08	0.94	1.78	13,170
17	0.1	73.3	9.2	2.9	0.09	1.04	1.77	13,200

*200 ml solution/30g of 32 mesh x 0 coal, 2 hrs at ambient temperature.

Treatment of Interior Basin coals with hydrogen peroxide/sulfuric acid resulted in 71-92% removal of the pyritic sulfur (as determined by differential pyritic sulfur analysis between treated and native coal), with heat content increases indicative of minimal reaction of hydrogen peroxide with coal (Table 9-6). A balanced equation for the reaction of hydrogen peroxide with pyrite (Eq. 7) shows that 3.75 moles of hydrogen peroxide are required for oxidation of each mole of pyritic sulfur removed from coal. This corresponds to about four weights of hydrogen peroxide per weight of pyritic sulfur removed from coal. Thus, for removal of 2% w/w pyritic sulfur from 1 ton of coal, 160 lbs of hydrogen peroxide would be required. Since a current price (13) for 50% w/w aqueous hydrogen peroxide is \$0.18/lb (in tank car quantities), a cost of nearly \$60 per ton of coal treated would be incurred for hydrogen peroxide consumption.

TABLE 9-6

Treatment of Three Interior Basin Coals with Hydrogen Peroxide – Sulfuric Acid*

Seam		Total	SO_4	Pyr	Org	Heat content Btu/lb
			Sulfur forms, % w/w			
Illinois No. 5	Native	3.6	0.08	1.06	2.46	12,590
	Treated	2.7	0.06	0.09	2.55	12,860
Iowa Bed	Native	8.7	2.00	3.93	2.77	10,700
	Treated	4.4	0.29	1.15	2.96	12,360
Oklahoma Bed	Native	4.1	0.48	2.05	1.57	13,300
	Treated	2.5	0.04	0.58	1.88	14,040

*250 ml solution/50g 32 mesh x 0 coal, ambient temperature for 1 hr with 15% w/w H_2O_2/0.3N H_2SO_4.

$$2 \, FeS_2 + 15 \, H_2O_2 \rightarrow Fe_2(SO_4)_3 + H_2SO_4 + 14 \, H_2O \qquad (7)$$

Ward (14) reports that analytical reagent hydrogen peroxide reacts vigorously and exothermically with coal. As soon as the slurry is warmed, an effervescence develops, apparently due to catalysis by coal constituents (probably FeS_2) of the breakdown of hydrogen peroxide to CO_2 and water. The reaction was thought by Ward to be dangerous and self-sustaining. If the coal-hydrogen peroxide slurry is heated for several days the organic matrix is completely destroyed leaving a pure residue suitable for analytical determination of coal ash.

It would appear that sulfuric acid and lower temperature stabilize the reagent used by Smith such that the spontaneous decomposition and reaction with coal is suppressed.

IV. CHLORINE

Sherman and Strickland (15) reported that dilute acidic aqueous chlorine solution reacts rapidly with pyrite, at or near room temperature, producing only ferric and sulfate ions (little or no elemental sulfur). Mukai *et al* (11) reported that treatment of several Japanese coals under similar conditions significantly reduces the pyritic sulfur content, but has an unfavorable effect on the caking properties of the coal. Since coal is highly susceptible to chlorination through aromatic substitution, this type of treatment must heavily chlorinate the coal. Unless chlorine can be removed from coal by subsequent steps, this type of treatment is probably not a desirable method for coal desulfurization.

V. POTASSIUM NITRATE

Given and Jones (16) conducted experimentation designed to remove pyritic sulfur from coal to prevent its retention in coke. They sought to accomplish this through the use of additives and through prior removal of pyritic sulfur with oxidizing or reducing ions, in combination with agents capable of complexing pyritic iron as it is dissolved.

Experiments with potassium nitrate in combination with the complexing agents, potassium cyanide and oxalic acid (Table 9-7), resulted in some removal of sulfur. This was apparently due to the oxidizing action of nitric acid which is in equilibrium with potassium nitrate in acidic solution. Calcium hypochlorite was used to treat coal at room temperature but only a slight oxidation of the pyritic sulfur was obtained.

TABLE 9-7

Treatment of a Pittsburgh Seam Coal with Aqueous
Potassium Nitrate Solutions

Experiment	Reagent	Conditions	Sulfur content* % w/w
	Native Coal		2.46
1	0.6M KCN 0.6M KNO$_3$	6 hrs @ reflux	1.90
2	0.04M Oxalic Acid 0.08M KNO$_3$	24 hrs @ RT, +8 hrs @ reflux	1.94

*1.95% w/w pyritic sulfur, 0.02% w/w sulfate sulfur, 0.49% w/w organic sulfur.

VI. REDUCTION

As shown in Chapter 3, the gas-solid hydrogenation of pyrite begins to be appreciable at about 450°C. Because this temperature is above the thermal decomposition range of coal and such reductions are the basis of synthetic liquid fuel production from coal, this approach will not be further discussed.

However, the reduction of coal pyrite by aqueous reducing agents at moderate temperatures would seem to be a method of desulfurization (or pyrite reaction) not yet reported in the literature. Several aqueous reducing agents were shown to have reduction potential sufficient to convert FeS_2 to $FeS+H_2$ (Chapter 3). These are: the metals Mg, Zn, Fe, Sn and Sn^{+2} and phosphorous acid (H_3PO_3). This latter is an example of a promising approach. Both H_3PO_3 and the product H_3PO_4 are water soluble and so should be washable from the coal matrix after reaction. The regeneration of H_3PO_3 could be accomplished by reduction of phosphoric acid or alternatively the oxidized acid would be sold and H_3PO_3 could be produced utilizing the credit. Any acid remaining on the coal would only increase the natural phosphorous content of coal.

Given and Jones (16) treated coal with stannous chloride and hydrochloric acid in an attempt to reduce pyrite to ferrous chloride and hydrogen sulfide and thereby partially desulfurize the coal. After 24 hours room temperature treatment and 8 hours at reflex essentially no desulfurization was obtained.

REFERENCES

1. R.A. Meyers, J.W. Hamersma and M.L. Kraft, Environ. Sci. and Tech., 9: (1) 71 (1975).

2. R.A. Meyers, J.W. Hamersma and M.L. Kraft, Unpublished experimentation (1972).

3. A.C. Elliott, Effluent Water Treat. J., 10: (7) 385 (1970).

4. R.J. Allison, P.H. Hatfield, R. Frumerman, U.S. Pat 3,495,945 (1970).

5. R.A. Meyers, J.W. Hamersma and M.L. Kraft, Unpublished experimentation (1972).

6. A.R. Powell and S.W. Parr, Univ. of Illinois Bull. No. 111, 62 (1919).

7. R.A. Mott, The Gas World — Coking Section (Supplement), 10 (July 1950).

8. H. Harms, Ger. Offen 1, 800,070 (1970).

9. Z.I. Syunyaev, R.N. Gimaev, Yu, M. Abyzgil'din, G.P. Malyatova and
    S.G. Zaitseva, Khim., Seraorg. Soedin. Soderzh. Neftyakh Nefteprod.,
    8, 381 (1968); Chemical Abstr. 71: 83257d (1969).

10. A.F. Diaz and E.D. Guth, U.S. Patent 3,909,211 (1975).

11. S. Mukai, Y. Araki, M. Konish and K. Otomura, Nenryo Kyoka:-sh:,
    48: (512) 905 (1969); Chem. Abstracts, 72: 123720d (1970).

12. E.B. Smith, Am. Chem. Soc. Dv. Fuel Chem. Preprints, 20: (2) 140
    (1975).

13. Chemical Marketing Reporter, January 5, 1976, p. 55.

14. C.R. Ward, Fuel, 53: 220 (1974).

15. M.I. Sherman and J.D.H. Strickland, AIME Transactions, 209, 1386
    (1957).

16. P.H. Given and J.R. Jones, Am. Chem. Soc. Dv. Fuel Chem., 8: (3) 185
    (1964).

CHAPTER 10

ORGANIC SULFUR REMOVAL PROCESSES

 I. Solvent Extraction

 II. Reduction

 III. Oxidation

 IV. Displacement Methods
     A. Alkali
     B. Acid

 V. Gas-Solid Processes

There are four methods for the removal of organic sulfur from coal which have been reported: solvent partition, reduction, oxidation and displacement. The following four sections present the results of liquid-solid processing involving these four methods, and the final section describes gas-solid treatment for oxidation, reduction and pyrolysis.

I. SOLVENT EXTRACTION

As discussed in Chapter 3, phenol has been found to extract significant quantities of organic sulfur from coal at temperatures near the solvent boiling point. However, since the researchers did not obtain a material balance for this experimentation, the selectivity of the dissolution is not known. In fact, it is possible that the coal extract and residue contain the same amounts of organic sulfur.

A systematic investigation of the potential of various solvents to dissolve organic sulfur compounds from coal was performed under the auspices of the Environmental Protection Agency. Initial studies (1) concentrated on screening a wide variety of solvents for their ability to partition organic sulfur from coal. These solvents included: 1) organic acids, 2) organic bases, 3) neutral organic solvents, 4) aqueous inorganic acids, and 5) aqueous caustic solutions. The results of the caustic solution extractions will be discussed in Section IV.

The screening study results for organic acids (phenols, nitrobenzene, formic acid), organic bases (pyridine, aniline) and neutral organic solvents (1-methyl-2-pyrollidone, dimethyl formamide) are summarized in Table 10-1. The analytical method consisted of simply measuring the total sulfur content of coal before and after treatment. Thus, the results could be in error if: a) any inorganic sulfur was removed or b) significant quantities of solvent were retained by the coal (even after drying in a vacuum oven to constant weight). The former was shown not to take place and the latter occurred to an extent defined further along in this discussion. Nevertheless, the method appears to be a reasonable screening device.

TABLE 10-1

Screening of Solvents for Organic Sulfur Removal*

Solvent	Extraction** T, $^{\circ}$C	Organic sulfur removal*** % w/w
o-chlorophenol	175	28
phenol	180	20
p-cresol	200	47
nitrobenzene	210	20
formic acid	100	0
pyridine	115	0
aniline	180	0
1-methyl-2-pyrollidone	200	0
dimethyl formamide	150	0

*Indiana No. 5 seam coal, 100 mesh x 0.

**3 hr extraction time.

***All residues were dried in vacuum oven to constant weight, some were post-extracted with water or methanol.

The results indicate that the phenols (very weak organic acids) and nitro-benzene (an electron acceptor) could potentially dissolve organic sulfur compounds from the coal matrix, but that the strong organic acid (formic acid), organic bases and neutral organics appear to have no special affinity for the organic sulfur portion of coal. Similar results were obtained for extraction of Pittsburgh, Illinois No. 6 and Bevier seam samples and for both longer and shorter extraction durations (1-24 hrs). Aqueous hydrochloric, hydrobromic, phosphoric, hexafluorophosphoric, sulfuric and nitric acids were also evaluated, showing no special ability to decrease the organic sulfur content of the coal residue. Para-cresol was selected as the most promising candidate extraction solvent because of its apparent high degree of organic sulfur partition, its low cost, the fact that it is a coal derived solvent, and its relative thermo-oxiditive stability (for recycle).

TABLE 10-2

P-Cresol Extraction: Screening of Coals for Organic Sulfur Removal*,**

Coal mine	Seam	Organic sulfur change, % w/w
Eagle No. 2	Illinois No. 5	-27 ± 12
Camp Nos. 1 and 2	No. 9	-23 ± 13
Orient No. 6	Herrin No. 6	$+31 \pm 33$
Belle Ayr	Roland-Smith	-12 ± 18
Fox	Lower Kittanning	-53 ± 12
Jane	Lower Freeport	-20 ± 18
Colstrip	Rosebud	-17 ± 9
	Ave	-17 ± 18

*1-1/2-hr extraction of -100 mesh coal at $200^{\circ}C$.

**All residues were washed with cold p-cresol then dried in a vacuum oven to constant weight (at least 36 hrs).

A screening (2) of coals was next performed for extraction of organic sulfur by p-cresol (Table 10-2). The coals varied from Appalachian (Fox and Jane mines) through Eastern Interior Basin (Eagle No. 2, Camp Nos. 1 and 2, and Orient No. 6) to Western (Belle Ayr and Colstrip). Here, as a refinement, organic sulfur removal was measured as the dry mineral matter-free (maf) organic sulfur change from starting coal to residue coal. The results indicated that the apparent organic sulfur partition between residue and solute can go both ways. That is, for the Orient No. 6, there was an apparent organic sulfur increase in the residue, while for the Fox mine coal there was a large decrease in the residue organic sulfur. Analysis of the extracts after removal of solvent show that: a) the extract was almost entirely organic in nature containing only 0.5 to 0.26% ash, and b) the maf sulfur content of the extract was essentially identical to the organic sulfur content of the starting coal with the exception of the Fox mine extraction where the sulfur content of the extract was approximately twice the organic sulfur content of the starting unextracted coal. Therefore, it would appear that only the Fox coal (from the Lower Kittanning seam) might be a viable candidate for desulfurization under these conditions.

Completely material balanced studies (3) were next performed for p-cresol extraction of Illinois No. 5 and Lower Kittanning seam coal samples (Table 10-3). For extraction, filtration and drying operations, all gases, solids and liquids were collected. The organic sulfur removal was estimated in three ways: a) recovered sulfur in the residue as compared with the sulfur recovery in the extract, recovered solvent and trapped gases, b) by the maf total sulfur content of the extracted coal, and c) the maf organic sulfur content of the residual coal. As before, analysis methods based on the extracted coal indicated that organic sulfur had been selectively dissolved from the coal matrix. However, analysis of recovered sulfur showed much smaller removals of organic sulfur. All results were corrected for solvent retention on the coal. This varied from 6 to 12% for the Illinois No. 5 and from 1.5 to 3.0% for the Lower Kittanning. Analyses based on the recovered sulfur content are the most accurate because they fully trace the sulfur "in" to the process and the sulfur "out" of the process. Roughly 8% by-weight of the feed coal is dissolved during the organic sulfur extraction under these conditions.

TABLE 10-3

P-Cresol Extraction: Material Balanced Studies*,** (3)

Coal seam	Run No.	Estimated organic sulfur removal % w/w		
		Recovered sulfur	Total sulfur content of extracted coal	Organic sulfur content of extracted coal
Illinois No. 5	1	12.4 ± 1.0	37 ± 6	33
	2	11.8 ± 0.8	46 ± 6	43
	3	6.0 ± 0.5	12 ± 4	10
	4	6.2 ± 0.5	17 ± 5	19
Lower Kittanning	1	7.7 ± 0.5	20 ± 20	–
	2	1.5	46 ± 20	–
	3	1.6	28 ± 20	–
	4	0	57 ± 20	–

*1 hr extraction of -14 or -100 mesh coal at 200°C.

** Solvent recovery was affected by vacuum drying.

Clearly, the simple extraction of native coal with organic solvent is not likely to lead to a viable method for the partition of organic sulfur compounds from coal. However, the effectiveness of the various depolymerization techniques cited in Chapter 3 should be investigated. Potentially, a small amount of p-toluene sulfonic acid or boron trifluoride etherate could be added to the various organic solvents to increase depolymerization and possibly increase selectivity. Also, extraction of depolymerized coal with solvents for removal of organic sulfur compounds from petroleum fractions (HF, H_2SO_4 or SO_2) might prove effective. Preheating of coal in an inert atmosphere at temperatures of $180\text{-}400^{\circ}C$ or partial hydrogenation at similar temperatures could increase the extract yield and possibly increase the selectivity. Other experimentation could concentrate on dissolving more of the coal matrix through increasing the temperature of extraction.

II. REDUCTION

Winkler (4) noted that metals such as iron, nickel, copper, tin, etc. scavenge organically bonded sulfur from hydrocarbon-type lubricating oils under friction conditions giving a sludge rich in metallic sulfides. It occurred to him that this reduction reaction might be useful for the removal of organic sulfur from coal. Tin-plated activated iron powder was contacted with coal in various oil media in a ball mill at $250^{\circ}C$ for 5 minutes. It is claimed that 53% of the total sulfur is removed from a Pittsburgh seam coal containing 1.5% organic sulfur and 0.2% pyritic sulfur. Non-activated iron powder and simple heating without ball milling gave lesser amounts of organic sulfur removal.

The process steps utilized were as follows: 1) grind coal, 2) mix ground coal with No. 6 fuel oil and iron powder to make a paste, 3) ball-mill the paste at $250^{\circ}C$, 4) dilute the paste with a solvent (tetralin), 5) centrifuge the mixture causing the higher density iron sulfide to be removed from the lower density coal and oil.

No cost estimates nor reagent regeneration methods were given for this quite imaginative approach to desulfurization and apparently no further work has been published.

III. OXIDATION

Friedman, LaCount and Warzinski (6) recently reported that a wide variety of coals may be treated with air in aqueous solutions at temperatures of 150-200°C and pressures of 500-1500 psi, to give at least 90% removal of pyritic sulfur (see Chapter 6) and up to 50% removal of organic sulfur. They believe that the organic sulfur which is removed is converted to sulfuric acid. The results shown in Table 10-4 indicate that the organic sulfur content of the coal residue is highly variable depending on the coal and temperature in some as yet uncorrelated way. The amount of organic sulfur removal (based on differential sulfur analysis of the starting coal and coal residue) varies from near zero (within analytical error) for the Mammoth, Wyoming No. 9, Lower Freeport and Minshall coals to vary substantial for the Pittsburgh and Illinois No. 6 coals.

TABLE 10-4

Organic Sulfur Removal by Air-Stream Process*

Coal (seam)	State	Temp, °C	Organic sulfur content, % w/w	
			Untreated	Treated
Bevier	Kans	150	2.0	1.6
Mammoth	Mont	150	0.5	0.4
Wyoming No. 9	Wyo	150	1.1	0.8
Pittsburgh	Ohio	180	1.5	0.8
Lower Freeport	Pa	180	1.0	0.8
Illinois No. 6	Ill	200	2.3	1.3
Minshall	Ind	200	1.5	1.2

*Nominal conditions: 1000 psi total pressure, 1 hr leach time.

The reduction in total sulfur content of the coal residue obtained by this process is shown in Table 10-5. More than 50% reduction in total sulfur content was obtained for the Mammoth and Pittsburgh (Ohio) coals and substantial reductions were obtained for all other coals. The authors did not publish directly the heat content changes but rather chose to cite the sulfur content per unit heat content (last two columns of Table 10-5). It can be seen that two of the coals were reduced in sulfur content while retaining sufficient heat content to meet the New Source Performance Standard requirements of 0.6 lbs of sulfur per 10^6 Btu (the Mammoth and Pittsburgh of Pa coals). The remaining coals did not meet this standard even though both pyritic and organic sulfur was attacked. Clearly, a portion of the coal matrix is being "burned" under the vigorous oxidative conditions utilized. Burning of the coal matrix results in a decrease in the ratio of organic coal matrix to coal mineral (ash), thus causing a diluting effect on the analysis of sulfur in the coal residue (since pyritic sulfur is simultaneously removed). Some of the apparent organic sulfur decrease shown in Table 10-4 could be attributed to this.

TABLE 10-5

Reduction of Total Sulfur by Air-Stream Process*

Coal (seam)	State	Temp, °C	Total sulfur, % w/w		Sulfur, lb/10^6 Btu	
			Untreated	Treated	Untreated	Treated
Minshall	Ind	150	5.7	2.0	4.99	1.81
Illinois No. 5	Ill	150	3.3	2.0	2.64	1.75
Lovilia No. 4	Iowa	150	5.9	1.4	5.38	1.42
Mammoth	Mont	150	1.1	0.6	0.91	0.52
Pittsburgh	Pa	150	1.3	0.8	0.92	0.60
Wyoming No. 9	Wyo	150	1.8	0.9	1.41	0.78
Pittsburgh	Ohio	160	3.0	1.4	2.34	1.15
Upper Freeport	Pa	160	2.1	0.9	1.89	0.80

*Nominal conditions: 1000 psi total pressure, 1 hr leach time.

TABLE 10-6

Calculated Heat Content Change* for Air-Stream Process

Coal (seam)	Heat content, Btu/lb**		Heat content change, %
	Initial	Treated	
Minshall	11,400	11,100	-3
Illinois No. 5	12,500	11,400	-9
Lovilia No. 4	11,000	9,900	-10
Mammoth**	12,900	11,500	-11
Pittsburgh (Pa)	14,100	13,300	-6
Wyoming No. 9**	12,800	11,500	-10
Pittsburgh (Ohio)	12,800	12,200	-5
Upper Freeport	11,100	11,300	+2

*Calculated from data in Table 10-5.

**Coal heat content $= \dfrac{\% \text{ sulfur in coal}/100}{\text{lbs S}/10^6 \text{ Btu}} \times 10^6$.

Calculations of the coal heat content before and after treatment (Table 10-6) shows that significant heat content decreases are almost always obtained on application of the air-steam process. With the exception of the Upper Free-port example, there is substantial combination of oxygen and/or loss of hydrogen resulting in heat content decrease in the residue coal. Also, significant formation of carbon monoxide and carbon dioxide volatiles must also take place (see Chapter 6).

IV. DISPLACEMENT METHODS

The reaction of caustic with coal to remove sulfur has been extensively studied and is proposed as an economical commercial method for coal desulfurization (7-10). The process developers assert that significant quantities of organic sulfur are removed (along with generally larger quantities of pyritic sulfur) during the caustic treatment, although there is reason for some skepticism (11). Acidic hydrolysis of sulfur containing groups in coal has also been reported (12). Both alkali and acid hydrolysis for removal of organic sulfur from coal are discussed in the following two sections.

A. Alkali

Masciantonio (7,8) investigated the treatment of bituminous coal with molten caustic (e.g., sodium and potassium hydroxide or mixtures of sodium, potassium and calcium hydroxide at temperatures of 150-400°C). Apparent organic sulfur removal from a Pittsburgh seam coal becomes appreciable at temperatures over 200°C, as shown in Figure 10-1. The overall process scheme (which is discussed in more detail in Chapter 8, with emphasis on pyritic sulfur removal) involves the treatment of pulverized (40 mesh x 0) coal with molten mixtures of potassium and sodium hydroxide and sometimes lime, decantation of the molten salt upper layer from the coal particle lower layer, washing of the coal with water and recycle of the molten caustic until its capacity is exhausted. It is postulated that the organic sulfur which is removed is converted to an alkali metal sulfide which dissolves in the fused salt. Higher temperatures (Figure 10-2) and extraction time of greater than 30 minutes result in maximum organic sulfur removal.

It should be pointed out that the organic sulfur removal data is based simply on the differential sulfur content between treated coal and residue. This method is subject to error (on the high removal side) provided that significant quantities of alkali are retained by the coal after washing. It has been noted that clay materials in coal can react with alkali to give insoluble sodium aluminum silicates (11) which if formed in large amounts would tend to give the residue a spurious low total sulfur content, indicative of high organic sulfur removal. Pittsburgh seam coal yield after extraction is reportedly 89-93% but no information is given as to the method of determination of the coal yield and whether yield could include alkali aluminum silicate.

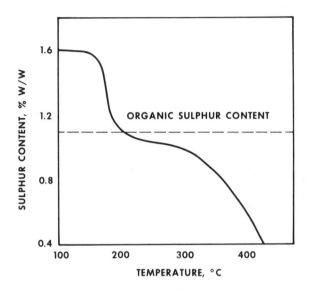

Figure 10-1. Organic Sulfur Content as a Function of Temperature,
Pittsburgh Seam Coal (7,8)

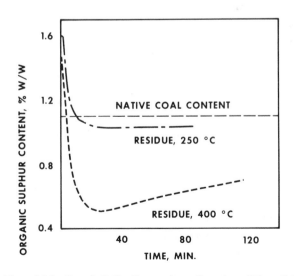

Figure 10-2. Organic Sulfur Removal as a Function of Time (7,8)

Masciantonio performed the most of his work on a relatively high rank Pittsburgh seam coal because Eastern Interior Basin and Western coals experienced a much greater degree of hydrolysis in the molten bath treatment. Only 69-76% yield of an Illinois coal and 52% yield of a Wyoming coal were recovered from the molten caustic treatment. Clearly, this degree of hydrolysis would be an economic disadvantage as the molten caustic would become fouled with large amounts of coal decomposition products. Hydrolysis of Wyoming subbituminous coal (13) with aqueous sodium hydroxide at temperatures of 200-425°C results in a high conversion to alkali soluble organics and was proposed as a method of converting coal to simple compounds.

Reggel, Raymond, Wender and Blaustein (11) report the desulfurization of Eastern Interior Basin coal by treatment with 10% aqueous sodium hydroxide (i.e., 2.5M). The pyritic sulfur content of the coals investigated was largely removed (these results are reported in Chapter 8). In general, the following procedure was followed: 1) pulverized coal was mixed with aqueous sodium hydroxide, 2) the mixture was heated in an autoclave to temperature, 3) on cooling, the slurry was acidified with carbon dioxide, hydrochloride acid, sulfur dioxide or sulfuric acid, 4) the coal residue (normally 91.5% maf basis) was obtained by filtration, and 5) the coal was dried. Organic sulfur removal was assessed on the basis of maf organic sulfur differential between the native coal and the residue. The results of sodium hydroxide and in one case calcium hydroxide extraction of two Illinois No. 6 coals are shown in Table 10-7. It can be seen that in no case is organic sulfur removal indicated; in fact, it appears that the organic sulfur content of the residue coal is sometimes "erratically increased" rather than decreased.

It is possible that organic sulfur compounds could be converted to sodium sulfide by action of sodium hydroxide (see Chapter 3). Acidification of the slurry would have the effect of releasing any sulfur, which has been removed from coal as sodium sulfide, in the form of hydrogen sulfide (Eq. 1). However, acidification would also neutralize sodium salts of organic compounds (Eq. 2,3) giving insoluble organics which would tend to be absorbed by the coal and not be removed by filtration. To the extent that the sodium salts of organic sulfur compounds are neutralized in this manner, potential organic sulfur removal would be depressed. The authors attempted to filter the aqueous caustic solution from the coal and then neutralize in order to eliminate possible deposition but the erratic organic sulfur increase still occurred.

It would thus appear that, under the conditions utilized, or for the coals investigated, a portion of the organic matrix lean in organic sulfur may sometimes be dissolved in preference to an even partition of organic sulfur compounds between coal residue and the alkali solution.

TABLE 10-7

Organic Sulfur Changes on Treatment of Coal
with Aqueous Sodium Hydroxide*

Mine**	Treatment	Workup	Ash % w/w	Organic sulfur % w/w, maf
River King	Native Coal		9.8	2.19
	NaOH	CO_2	12.4	2.06
	$Ca(OH)_2$	HCl	8.16	2.14
	NaOH	HCl	0.67	2.44
	H_2O	CO_2	9.46	1.96
	H_2O	HCl	8.76	1.98
	NaOH	SO_2	0.72	1.99
	NaOH	H_2SO_4	0.52	2.33
Elliot	Native Coal		18.15	1.05
	NaOH	CO_2	22.84	2.44
	NaOH	HCl	5.11	1.75
	NaOH, 325°C	HCl	8.43	2.46
	NaOH	H_2SO_4	7.26	1.94

*10% aqueous NaOH for 2 hrs at 225°C in an autoclave followed by acidification (workup).

**200 mesh x 0.

$$Na_2 S + CO_2 + H_2O \rightarrow Na_2 CO_3 + H_2S \tag{1}$$

$$2 R S Na + H_2O + CO_2 \rightarrow Na_2 CO_3 + 2 R S H \tag{2}$$

$$2 \underset{R}{\underset{\bigcirc}{}}^{ONa} + H_2O + CO_2 \rightarrow 2 \underset{R}{\underset{\bigcirc}{}}^{OH} + Na_2 CO_3 \tag{3}$$

An Indiana No. V coal was extracted with 20% aqueous sodium hydroxide (5M) for 3 hrs at 115°C (1). The caustic solution was filtered away from the extracted coal with some difficulty as the coal had been heavily degraded. The coal was washed several times with fresh water. Assessment of organic sulfur removal, which was made on the basis of differential organic sulfur content between native and residue coal, indicated an apparent 31% organic sulfur removal. These results, being a single point, may well also be subject to the erratic behavior noted by Reggel *et al.*

Extensive investigations have been performed at the Battelle Memorial Institute on the removal of both pyritic and organic sulfur by aqueous caustic solutions (9,10). Pyritic sulfur removal results and a consideration of engineering factors are presented in Chapter 8. The optimum experimental conditions for removal of sulfur from coal involves treatment with an aqueous solution containing 10% w/w sodium hydroxide and about 2% calcium hydroxide at temperatures between 225 and 350°C (giving rise to pressure of 350 to 12,500 psi) for periods to 30 minutes. The extraction solution was separated from the residue by centrifugation. These conditions would appear to be very similar to those utilized by Reggel *et al* (11), with the exception of the addition of lime. A summary of experimental results for removal of organic sulfur from several mines is shown in Table 10-8. Up to 72% of the organic sulfur appears to be removed when the maf organic sulfur content of the native coal is compared to the organic sulfur content of the treated coal. It is reported (10) that only about 5% of the coal is dissolved by this treatment, which is considerably less than that reported by either Masciantonio (7,8) or Reggel *et al* (11).

TABLE 10-8

Extraction of Organic Sulfur with Aqueous Alkali

Experiment	Coal mine	Seam	Organic sulfur content maf % w/w		Organic sulfur removal % w/w
			Initial	Treated	
1	Sunny Hill	No. 6	1.1	0.6	41
2	Martinka No. 1	Lower Kittanning	0.7	0.5	24
3	Montour No. 4	Pittsburgh	1.0	0.3	72
4	Westland	No. 8	0.8	0.5	38
5	Beach Bottom	No. 8	1.0	0.7	30
6	Meigs No. 1	Clarion 4A	2.3	1.1	52

Kasehagen (14) has studied the action of aqueous sodium hydroxide for 20-30 hrs on a Pittsburgh seam coal over the temperature range of 250-400°C and at various alkali concentrations, including those cited in the above references. He reports the decomposition of the Pittsburgh seam coal into a coke-like residue and phenols, acids, neutral oils, hydrocarbon gases and carbon dioxide. Variable amounts of products were obtained at 330°C as a function of alkali concentration (Figure 10-3) and temperature (with 5N sodium hydroxide) (Figure 10-4). It can be seen that 10% sodium hydroxide (2.5N) treatment of Pittsburgh seam coal at 250-300°C, which corresponds to the thermal and chemical conditions and coal seam reported by the Battelle researchers (for Experiment 3 of Table 10-8), should give rise to significant amounts of phenols, acids, carbon dioxide, hydrocarbon gases and some neutral oils although lower yields would be obtained at the 30 minute residence time. The alkali-soluble material was shown by Kasehagen to be predominately phenolic (after releasing such compounds from the alkaline solution with carbon dioxide). Kasehagen also performed alkaline hydrolysis of Illinois No. 6 coal (corresponds to Experiment No. 1 of Table 10-4) which showed over 30% of the carbon to be converted to phenolic and acidic products compared with about 12% for the Pittsburgh seam under the same conditions. Phenolics or organic acids (Figure 10-5) have been reported from the alkali treatment of various types of coal and coal extracts (13). Pure carbon reacts with sodium hydroxide (16) to form sodium carbonate and hydrogen (Eq. 4).

$$C + 2\,Na\,OH + H_2O \;\rightarrow\; Na_2\,CO_3 + 2\,H_2 \qquad\qquad (4)$$

Kasehagen also noted, in most cases, a decrease in organic sulfur (Table 10-9) in the residues. However, some runs at high caustic concentration revealed an increased organic sulfur content. The measured nitrogen content of the residue tended to increase, but this was thought to be due to a bias in the method of analysis.

It has been proposed (10) that spent alkali be regenerated by treatment with carbon dioxide to form hydrogen sulfide and sodium carbonate as in Eq. 1 and that sodium carbonate can be converted by treatment with lime to sodium hydroxide and insoluble calcium carbonate product. This first step in regeneration would result in formation of insoluble solid and liquid organic phenols, acids and possibly thiols (see Eqs. 2 and 3) which must be removed by either filtration or decantation.

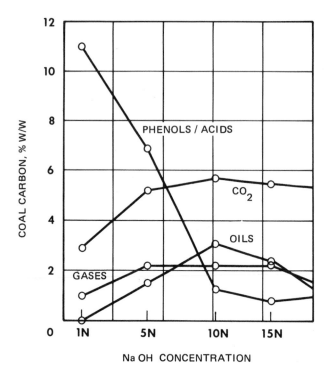

Figure 10-3. Product Yield as a Function of Alkali Concentration Using
Pittsburgh Seam Coal (14)

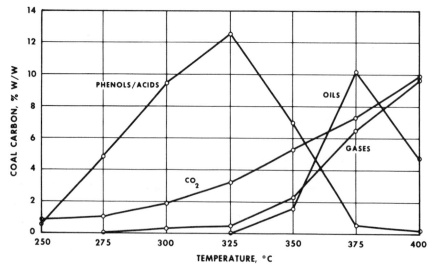

Figure 10-4. Product Yield as a Function of Temperature Using
Pittsburgh Seam Coal (14)

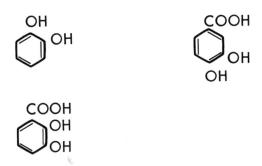

Figure 10-5. Coal Hydrolysis Products

TABLE 10-9

Extraction of Coal with Aqueous Alkali (14)

NaOH		% w/w*			
Temp °C	Concn.	C	H	N	S**
250	5N	84.15	5.96	2.16	0.42
275	5N	86.30	5.61	2.15	0.32
300	5N	87.15	5.53	1.66	0.34
325	5N	88.95	5.27	2.12	0.15
325	60%	88.11	5.66	2.26	0.26
350	1N	89.38	4.78	2.36	0.16
350	5N	90.43	4.48	2.05	0.12
350	10N	90.39	4.78	0.98	0.21
350	15N	90.66	5.04	1.94	0.10
350	60%	90.16	5.59	1.55	0.42
350	80%	89.26	5.48	1.64	0.73
350	100%	85.85	5.44	1.72	0.78
375	5N	92.01	3.85	1.83	0.22
400	5N	92.93	3.62	1.82	0.13
400	60%	92.04	4.36	1.68	0.58
Original coal		84.99	5.68	1.66	0.65

*maf basis.

**Organic sulfur only.

Alternatively, it might be possible to regenerate the alkali in a single step by oxidation with air or oxygen (Eq. 5, 6) and then decant or filter the oxidized organic products from the regenerated solution.

$$2 \, R \, S \, Na + H_2O + \frac{1}{2} O_2 \rightarrow R \, S \, S \, R + 2 \, Na \, OH \tag{5}$$

$$2 \, \underset{R}{\bigcirc}{}^{O\,Na} + O_2 \rightarrow O = \underset{R}{\bigcirc}{}^{R} = \bigcirc = O + 2 \, Na \, OH \tag{6}$$

B. Acid

Until very recently there has been no data which indicates any hydrolytic breakdown of the coal matrix in other than alkaline medium. In fact, hydrochloric or sulfuric acids appear to be without hydrolytic effect at least up to temperatures which can be reached without pressure equipment (17). However, coal has been treated with boiling hydriodic acid at atmospheric pressure (12) with special attention to the behavior of organic sulfur compounds. It was found that thiol groups split off hydrogen sulfide on treatment with HI, and this method was used for determination of thiol groups in various ranks of coal.

V. GAS-SOLID PROCESSES

Block, Sharp and Darlarge (18), report the partial removal of organic and inorganic sulfur from several coals by gas treatment in a stationary bed. The inorganic sulfur results are described in Chapter 7.

Hydrogen, steam and air were evaluated and found to be most effective for removal of both types of sulfur at temperatures of 900, 600 and 450°C, respectively. Averages of 76%, 25% and 20% removal of organic sulfur were found for hydrogen, steam and air treatment at the optimum conditions for 10 U.S. high-volatile bituminous coals. Some typical results are summarized in Table 10-10. Organic sulfur content was determined as the difference between the coal total sulfur content and the sulfur content of the coal residue after extraction with nitric acid (to remove pyritic and sulfate sulfur and presumably any sulfide sulfur formed by pyrolysis or reduction of pyrite).

TABLE 10-10

Desulfurization by Stationary Bed Gas Treatment

Coal seam	Treatment	Organic sulfur, % w/w		Wt loss % w/w
		Untreated	Residue	
Pittsburgh No. 8	Hydrogen	2.09	0.31	38.1
	Steam	2.09	1.55	26.7
	Air	2.09	1.71	34.0
Illinois No. 6	Hydrogen	1.78	0.48	36.3
	Steam	1.78	1.25	24.7
	Air	1.78	1.86	18.6

The results showed a slight selectivity for volatilization of organic sulfur as a result of pyrolysis and oxidation by air. However, fairly severe coal loss as oxidized volatile material (mostly CO_2) was encountered, and the heat content of the remaining char was reduced by about 10%. Similar results were found for steam treatment but there was less residue heat content loss. Calculation of organic sulfur removal based on the difference in organic sulfur between native coal and residue is subject to error (on the high side) as the sulfur content of the residue is diluted by increased ash and oxygen. Hydrogen treatment is indeed effective in removing organic sulfur, while maintaining the fuel value of the residue, through addition of hydrogen to the coal matrix. However, a large penalty in volatilized coal is seen because of the high temperature utilized. Some engineering and cost considerations for this type of desulfurization are discussed in Chapter 7.

REFERENCES

1. R.A. Meyers, J.S. Land and C.A. Flegal, Final Report Chemical Removal of Nitrogen and Organic Sulfur from Coal, Air Pollution Control Office Contract No. EHSD 71-7, May 1971.

2. J.W. Hamersma, M.L. Kraft, C.A. Flegal, A.A. Lee and R.A. Meyers, Applicability of the Meyers Process for Chemical Desulfurization of Coal, EPA-650/2-74-025, U.S. Environmental Protection Agency, Washington, D.C., April, 1974.

3. J.W. Hamersma, E.P. Koutsoukos, M.L. Kraft, R.A. Meyers, G.J. Ogle and L.J. Van Nice, Chemical Desulfurization of Coal: Report of Bench-Scale Developments Volume 1, EPA-R2-73-173a, U.S. Environmental Protection Agency, Washington, D.C., February, 1973.

4. J. Winkler,·Am. Chem. Soc. Dv. Fuel Chem. Preprints, 12: (4) 19 (1968).

5. A. Gergina and V. Kovacheva, Khim. Tverd. Topl. 1969 (5) 22, Chem. Abstracts 72: 23245j.

6. S. Friedman, R.B. LaCount, and R.P. Warzinski, to be published in Am. Chem. Soc. Dv. Fuel Chem. Preprints, 22 (1977).

7. P.X. Masciantonio, U.S. Patent 3,166,483 (1965).

8. P.X. Masciantonio, Fuel, 44: 269 (1965).

9. E.P. Stambaugh, J.F. Miller, S.S. Tam, S.P. Chauhan, H.F. Feldmann, H.E. Carleton, J.F. Foster, H. Nack, and J.H. Oxley, Hydrocarbon Processing, 115 (July 1975).

10. Science, 189: 129 (1975).

11. L. Reggel, R. Raymond, I. Wender and B.D. Blaustein, Am. Chem. Soc. Div. Fuel Chemistry Preprints, 17: (1), 44 (1972).

12. G. Angelova and V. Kovacheva, Khim Tverd. Tupl., 1969 (5) 22: Chem. Abstracts 72:23245j.

13. F.G. Parker, J.P. Fugassi and H.C. Howard, Industrial and Engineering Chemistry, 47: (8), 1586 (1955).

14. L. Kasehagen, Ind. Eng. Chem., 29: 600 (1937).

15. H. Strache and R. Lant, Kohlenchemie Akaclemishe Verlagsgesellschaft, Leipzig, 286 (1924).

16. H.H. Lowry chrm., Chemistry of Coal Utilization Vol. I, John Wiley and Sons, Inc., New York, 1945 pp 418-423.

17. F. Haber and L. Bruner, Z. Electrochem, 10: 697 (1904).

18. S. Block, J.B. Sharp and L. Darlage, Fuel, 54: 113 (1975).

AUTHOR INDEX

A

Abramov, A.A.	39, 40
Agarwal, J.C.	185, 187
Allen, E.T.	30
Allison, R.J.	211
Amyzgil'din, Yu M.	216
Angelova, G.	230, 238
Araki, Y.	216, 220
Averitt, P.	3

B

Baker, R.D.	99
Baldwin, R.M.	96, 191
Baranskii, A.D.	46
Bartholomew, F.J.	76, 101
Biernat, R.J.	27, 28
Birch, S.F.	49
Blaustein, B.D.	204, 205, 230, 232
Block, S.S.	194, 199, 238
Blum, I.	191
Brooks, J.D.	21
Bruner, L.	238

C

Carleton, H.F.	205, 206, 230, 234
Cernivec, S.	102
Chapman, W.	23
Chatterjee, A.K.	49
Chertkov, Y.B.	47
Chauhan, S.P.	205, 206, 230, 234
Cindia, V.	191
Clark, C.S.	23
Crenshaw, J.L.	30

D

Darlage, L.J.	194, 199, 238
Deurbrouck, A.W.	4, 7, 11, 12, 13
Diaz, A.F.	216
Dryden, I.G.C.	47

E

Elliott, A.C.	211
Emery, J.E.	3
Erdman, J.G.	50
Evering, B.L.	46
Emilov, V.V.	62

F

Feldman, H.F.	205, 206, 230, 234
Finkelstein, M.	30, 50
Fisenko, N.N.	46
Flegal, C.A.	78, 223, 225, 234
Foerster, F.	4
Foster, J.F.	205, 206, 230, 234
Friedman, S.	184, 228
Frumerman, R.	211
Fugassi, J.P.	232, 235
Fujimori, T.	76

G

Garrels, R.M.	31, 40
Gary, J.K.	62
Gary, J.H.	96, 191
Geer, M.R.	4
Geisler, W.	4

Giberta, R.A.   181, 182,
                         184, 185,
                         186, 187   Ichikuni, M.   181
Gimaev, R.N.   216   Imuta, K.   46, 51
Given, P.H.   17, 18, 23,   Irminger, P.F.   181, 182,
                         42, 220,   184, 185,
                         221   186, 187
Gluskoter, H.J.   4   Ishikawa, K.   76
Golden, J.O.   96, 191
Gould, E.S.   28, 30   J
Gowan, W.S.   49
Gruse, W.A.   47, 48   Jacobs, T.K.   193, 199
Guth, E.D.   216   Jones, J.R.   220, 221

                H   K

                                     Kasehagen, L.   204, 207,
Haber, F.   238   208, 235,
Halpern, J.   38, 40, 41,   236, 237
                        183   Kendrick, W.P.   61
Hamersma, J.W.   4, 6, 22,   Kennedy, T.H.   9, 10
                        27, 29, 31, Keyes, D.B.   178
                        40, 60, 61, Kharasch, N.   42, 49
                        62, 63, 66, Kimberlin, C.N., Jr. 46
                        67, 68, 70, Konish, M.   216, 220
                        71, 76, 78, Koutsoukos, E.P.   31, 40, 61,
                        79, 81, 83,   62, 63, 66,
                        96, 97, 98,   67, 68, 70,
                        99, 102,   71, 74, 76,
                        103, 104,   78, 79, 80,
                        108, 155,   81, 83, 85,
                        156, 161,   87, 89, 90,
                        164, 180,   91, 93, 95,
                        191, 211,   97, 98, 99,
                        212, 225,   104, 106,
                        226   112, 115,
Handwerk, J.G.   96, 191   116, 117,
Harms, H.   215, 216   118, 122,
Hatfield, P.H.   211   123, 125,
Haver, F.P.   30, 99   137, 139,
Hensley, E.F.   115, 116,   141, 142,
                        155, 157,   145, 149,
                        158, 159,   150, 155,
                        160   156, 158
Heredy, L.A.   46   164, 182,
Howard, H.C.   232, 235   184, 185,
                                                           226

Kovacheva, V. — 230, 238
Kraft, M.L. — 4, 6, 22, 27, 29, 31, 40, 60, 61, 62, 63, 66, 67, 68, 70, 71, 74, 76, 78, 79, 80, 81, 83, 85, 86, 87, 89, 90, 91, 93, 95, 97, 98, 99, 102, 103, 104, 106, 108, 115, 116, 118, 122, 123, 125, 137, 139, 141, 142, 145, 149, 150, 155, 156, 158, 161, 164, 211, 213, 225, 226
Kramarsic, V. — 102
Kunda, W. — 82, 106

L

Lacount, R.B. — 184, 228
Lahiri, A. — 49
Land, J.S. — 60, 61, 112, 223, 234
Larina, V.A. — 46
Latimer, W.M. — 28
Lee, A.A. — 78, 225
Leonard, J.W. — 20, 21, 122
Li, S.H. — 176
Lien, A.P. — 46
Liepna, L. — 62

Lissner, A. — 48
Lorenzi, L. — 61, 112
Lowry, H.H. — 17, 175, 235
Lyalikova, N.N. — 187, 189

M

Macejevskis, B. — 62
Maier, C.G. — 78
Majima, H. — 39, 40
Malyatova, G.P. — 216
Mark, H. — 98, 99
Masciantonio, P.X. — 201, 202, 203, 230, 231, 234
Mathews, C.T. — 32, 33, 39, 40, 78
Mazumdar, B.K. — 49, 61, 94
McCaulay, D.A. — 46
Magee, E.M. — 115, 164, 165
McKay, D.R. — 38, 40, 41, 183
Meadow, J.R. — 46
Mellors, J.W. — 28, 29, 30, 36, 38, 82, 94
Meyers, R.A. — 10, 27, 29, 31, 40, 60, 61, 62, 63, 66, 67, 68, 70, 71, 74, 76, 78, 80, 81, 83, 85, 86, 87, 89, 90, 91, 93, 94, 95, 96, 97, 98, 99, 104, 106, 112, 115, 116, 117, 118, 122, 123, 125,

137, 139,
141, 142,
145, 149,
150, 155,
156, 158,
164, 180,
182, 184,
185, 191,
211, 212,
223, 225,
226, 232

Miller, J.F.   205, 206,
230, 234

Mitchell, D.R.   20, 21,
122

Mirkus, J.D.   193, 199
Mott, R.A.   49, 215
Muntean, V.C.   191
Mukai, S.   216, 220

N

Nack, H.   205, 206,
230, 234

Nackiw, V.N.   82, 106
Nekervis, W.F.   115, 116,
155, 157,
158, 159,
160

Nelson, H.W.   178
Nelson, J.B.   20
Nemes, A.   48
Neuworth, M.B.   46
Nickless, G.   66
Nielsen, G.F.   3
Norris, P.   49

O

Ogle, G.J.   31, 40, 62,
67, 68, 70,
71, 76, 78,
79, 81, 83,
97, 98, 99,

155, 156,
164, 182,
184, 185,
226

Orsini, R.A.   74, 80, 83,
85, 86, 87,
89, 90, 91,
93, 95, 98,
99, 101,
104, 106,
115, 116,
117, 118,
122, 123,
125, 137,
139, 141,
142, 145,
149, 150,
158, 182,
184, 185

Ouchi, K.   46, 51
Oxley, J.H.   205, 206,
230

P

Parker, F.G.   232, 235
Parr, S.W.   175, 176,
215

Peters, E.   39, 40
Petrovic, L.J.   181, 182,
184, 185,
186, 187

Philinis, J.   36, 37, 40,
191

Pound, J.R.   76, 77
Powell, A.R.   45, 175,
215

R

Raymond, R.   204, 205,
230, 232

Reggel, L.   204, 205,
230, 232

Reid, E.E. — 42, 47, 48, 49

Rhys Jones, D. — 23

Robins, R.G. — 27, 28, 32, 33, 39, 40, 78

Rogoff, M.H. — 187, 188

Romanteev, Y.P. — 62

Ross, S.D. — 30, 50

Rudd, E.J. — 30, 50

Rudyk, B. — 82, 106

S

Sato, M. — 39, 40

Santy, M.J. — 61, 74, 80, 83, 85, 86, 87, 89, 90, 91, 93, 95, 98, 99, 104, 106, 115, 116, 117, 118, 122, 123, 125, 137, 139, 141, 142, 145, 149, 150, 158

Sareen, S.S. — 181, 182, 184, 185, 186, 187

Shchurouskii, Y.P. — 62

Schwab, G.M. — 36, 37, 40, 191

Senning, A. — 42

Sercombe, E.J. — 62

Sharp, J.B. — 194, 199, 238

Sherman, M.I. — 220

Shmuk, Ye I. — 187, 189

Silverman, M.P. — 187, 188

Simon, J.A. — 4

Singer, P.C. — 36, 82

Singer, S.F. — 1, 8

Sinha, R.K. — 191, 192, 196, 199

Sinke, G.C. — 116

Smith, E.B. — 217

Snow, R.D. — 178

Stambaugh, E.P. — 205, 206, 230, 234

Stevens, D.R. — 47, 48

Stokes, H.N. — 29, 33, 40

Strickland, J.D.H. — 220

Stumm, W. — 36, 82

Syunyaev, Z.I. — 216

T

Tam, S.S. — 205, 234

Tek, Rasin, M. — 108, 173

Thompson, M.E. — 31, 40

Thornton, A.W. — 9, 10

Trindade, S.C. — 4

V

Van Krevelen, D.W. — 17

Van Nice, L.J. — 31, 40, 61, 62, 67, 68, 70, 71, 74, 76, 78, 79, 80, 81, 83, 85, 86, 87, 89, 90, 91, 93, 95, 97, 98, 99, 104, 106, 112, 115, 116, 117, 118, 122, 123, 125, 137, 139, 141, 142, 145, 149, 150, 155, 156, 158, 164, 226

W

Walker, P.L.   191, 192,
                          196, 199
Ward, C.R.   219
Warren, I.H.   183
Warzinski, R.P.   184, 185,
                          228
Watanbe, A.   101
Wender, I.   187, 188,
                          204, 205,
                          230, 232
Wheeler, R.V.   45
White, T.A.   46

Winkler, J.   227
Wise, W.S.   42
Wong, M.M.   30
Woolhouse, T.G.   4
Wyss, W.F.   17, 42, 53

Y

Yamashita, Y.   46, 51
Yancy, H.F.   4

Z

Zaitseva, S.G.   216
Zarubina, Z.M.   187, 189

SUBJECT INDEX

A

Acetonitrile, 31, 40
Acid, effect of
on Fe^{+3} reaction with FeS_2, 30, 33
on H_2O_2 reaction with FeS_2, 217, 219
on Meyers Process, 62, 64, 67, 101, 103
on O_2 reaction with FeS_2, 39, 40
Africa, coal of, 2, 4
Air, reaction of,
with coal, 175, 183, 184, 229
with disulfides, 44
with ferrous salts, 61, 100-101, 115, 120-121
with FeS_2, 35-40, 76-94, 175-189, 191-195, 197-200
with sulfides, 44
with thiols, 44
America, Central, coal of, 2
America, South, coal of, 2, 4, 22
Amines, in coal, oxidation of, 50
Ammonia, as desulfurization media, 29, 30, 97
Ammonium polysulfide, 99-100
Aniline, 50, 182, 224
Anisole, 50
Anthracene, 50
Antimony,
in coal, 22
removal from coal, 102-106
Appalachian Basin coal,
in Meyers Process, 64, 107-114
sulfur content of, 5-6
Aromatic compounds,
in coal, 17-19
oxidation of, 50

Arsenic,
in coal, 22
removal from coal, 102-106

B

Base, desulfurization with,
of pyrite, 29
of pyrite in coal, 201-209
of organic sulfur compounds, 43, 48, 51, 230-238
Battelle Hydrothermal Process, 204-209, 234-238
Belgium, coal of, 2
Benzene, 50, 98
Beryllium,
in coal, 22
removal from coal, 102-106
Boron,
in coal, 22
removal from coal, 102-106
Boron trifluoride,
depolymerization of coal with, 46
desulfurization of coal with, 227
Brazil, coal of, 4
Briquetting of coal after Meyers Process, 160
Bureau of mines, research projects,
on caustic leaching, 204-206, 232-233
on hydrogen peroxide leaching, 216-219.
on air oxidation, 184-185, 228-229

C

Calcium,
in coal, 20

in Meyers Process, 103, 106
removal from coal, 103, 106-107
use in desulfurization, 230-238
Canada, coal of, 2, 4
Carbon disulfide as sulfur solvent, 98
Carbonyl groups in coal, 17-18
Catalysis in desulfurization
processes,
    for depolymerization, 43, 46-47,
    227
    for oxidation, 175
    for reagent regeneration, 76-77
    for reduction, 48
Cadmium,
    in coal, 22
    removal from coal, 102-106
Caustic (see sodium hydroxide)
Chemical constitution of coal, 17-20
China, coal of, 2-4
Chlorination of coal, 66
Chlorine,
    as residue in desulfurization, 66
    for desulfurization, 220
Cleaning of coal,
    compared to chemical desulfuri-
    zation, 10, 12, 108-109, 114-115
    for coke making, 9, 10
    for meeting air pollution stand-
    ards, 11-13, 108-109
    for trade element removal, 104-
    105
    effect on residual pyrite particle
    size, 23
    in Meyers Process,
        effect on cost, 115, 148-155
        effect on meeting standards,
        114-115
    effect on rate, 74-75, 148
Coal tar oil, 98
Costs of chemicals, 56
Costs of processes (see specific
processes)
Chromium,
    in coal, 22

removal from coal, 102-106
Copper,
    in coal, 22
    removal from coal, 102-106
Czechoslovakia, coal of, 2

D

Decomposition (see type of
reaction)
Desulfurization (see specific
chemicals)
Dimethylformamide, 31, 224
Dimethyl sulfoxide, 31
Displacement,
    of organic sulfur, 44, 51
    of organic sulfur in coal, 230-238
    of pyritic sulfur, 26
Disulfides,
    in coal, 18
    nucleophilic displacement of, 44,
    51
    oxidation of, 44
    reduction of, 43, 48
Dow Chemical — U.S.A., Meyers
Process design, 155-164

E

Eastern Interior Basin Coal,
    in Meyers Process, 110-112,
    sulfur content of, 6
Electrode potential,
    of aromatics in coal, 50
    of FeS_2, 28, 41
    of FeS_2 oxidants, 28
    of FeS_2 reductants, 41
    of solvents for desulfurization,
    31
Engineering design,
    of Air Oxidation Process, 197-
    198
    of Battelle Process, 208-209
    of Ledgemont Process, 185-186

of Meyers Process,
  by Dow Chemical — U.S.A., 155-164
  by Exxon, 164-169
  by TRW Inc., 115-154
  of Molten Caustic Process, 203
  of SO_2 process, 214
Exxon Research and Engineering Co., Meyers Process design, 164-169
Extraction, desulfurization by solvent, 42, 45-46, 223-226

                          F

Ferric alum, reaction with FeS_2, 30
Ferric ammonium citrate, reaction with FeS_2, 33
Ferric ammonium oxalate, reaction with FeS_2, 33
Ferric chloride
  in the Meyers Process, 62
  reaction with FeS_2, 27, 31,33, 63
Ferric sulfate,
  in the Meyers Process,
    as a product, 100-101
    reaction with FeS_2, 62-93
    regeneration of, 76
    selectivity of, 64, 67, 109-112
  cost of, 55-56
  in acid mine drainage, 36
  reaction with FeS_2, 30, 31-35
  role in oxygen processes, 35-36
Ferrobacillus ferrooxidans,
  in FeS_2 oxidation, 36
  in FeS_2 oxidation in coal, 187-189
Formic acid, 224
France, coal of, 2

                          G

Gas generation during desulfurization, 183-184
Gas-solid desulfurization processes, 191-199
Germany, coal of, 2-4

                          H

Hydrogenation of coal, 44, 196-197
Heat content,
  effect of Battelle Process on, 206-207
  effect of Air-Steam process on, 229-231
  effect of gas treatment on, 192-197
  effect of H_2O_2 on, 217-219
  effect of Meyers Process on, 109-111
  effect of SO_2 process on, 212-213
Hexamethyl phosphotriamide, 31
Hexane, 98
Hydrofluoric acid
  extraction of coal with, 47
  extraction of petroleum with, 46
Hydrogen Iodide, extraction of organic sulfur with, 51
Hydrogen Peroxide,
  electrode potential of, 28
  reaction with FeS_2, 28-29
  reaction with FeS_2 in coal, 216-219
Hydrogen, reaction with,
  coal, 44, 196-197
  elemental sulfur, 99
  FeS_2, 26, 40-41
  organic sulfur, 48-49
Hypochlorite oxidation of organic sulfur, 49

I

India, coal of, 2, 4
Iron,
  compounds,
    electrode potential of (see
    individual compounds)
    reaction with coal (see indi-
    vidual chemicals)
  forms,
    in coal, 21
    in Meyers Process leach solu-
    tion, 76, 100
  metal,
    reaction with FeS_2, 40-41
    reaction with organic sulfur,
    227
  pyrites,
    crystalline form, 21
    occurrence in coal, 3-5, 107
    physical structure, 21, 23
    reaction with chemicals (see
    individual chemicals)

J

Japan, coal of, 2, 4

K

KVB Process, 216-217

L

Lead,
  in coal, 22
  removal from coal, 102-106
Lead carbonate, 29
Ledgemont Process, 181-186
Lithium,
  in coal, 22
  removal from coal, 102-106

Low-temperature
  pyrolysis of organic sulfur compounds, 43, 47-48
  hydrogenation, 43-44

M

Magnesium, 40
Malaysia, coal of, 4
Mesitylene, 50
1-methyl-2-pyrollidone, 224
Meyers Process, 60-169
Magnesium,
  removal from coal, 106
Mexico, coal of, 2
Microstructure of coal, 17-23
Mine
  Beach Bottom, 206, 207, 234
  Belle Ayr, 22, 225
  Belmont, 206
  Bird No. 3, 22, 73, 109, 110,
  114, 115, 180
  Camp Nos. 1 and 2, 22, 109,
  110, 225
  Colstrip, 22, 109, 110, 225
  Dean, 73, 109, 110, 112
  Delmont, 22, 109, 110, 114,
  115
  Derbyshire, 4
  Eagle No. 1, 206
  Eagle No. 2, 4, 22, 109, 110,
  180, 225
  Egypt Valley, 109, 112
  Egypt Valley No. 21, 22, 110
  Elliot, 205, 233
  Fernie, 4
  Fox, 22, 73, 109, 110, 112, 225
  Harris Nos. 1 and 7, 73, 109,
  110, 112
  Homestead, 73, 109, 110
  Humphrey No. 7, 22, 109, 110,
  112

Isabella, 73, 109, 110, 180
Jane, 22, 109, 110, 112, 225
Ken, 22, 73, 109, 110, 206
Kopperston, 73, 109, 110, 112
Lower Newcastle, 4
Lucas, 22, 73, 109, 110, 114
Marion, 22, 73, 109, 110, 114
Martinka, 73, 110, 112, 114, 115, 204, 207
Martinka No. 1, 109, 204, 207, 234
Mathies, 22, 73, 108, 109, 110
Meigs, 22, 73, 110
Meigs No. 1, 109, 234
Miike, 4
Montour, 207
Montour No. 4, 206, 234
Muskingum, 22, 73, 109, 110
Navajo, 109, 110
North River, 73, 108, 109, 110, 114, 115
No. 1, 109, 110
Orient No. 6, 22, 109, 110, 180, 225
Powhattan No. 4, 22, 73, 109, 110
Renton, 206, 207
River King, 205, 233
Robinson, 22, 109, 110
Santa Caterina, 4
Sarawak, 4
Shakhtersky, 4
Shoemaker, 73, 109, 110
Star, 73, 109, 110
Sunny Hill, 206, 234
Taitung, 4
Tipong, 4
Transvaal, 4
Vermillion, 178
Warwick, 22, 109, 110
Weldon, 109, 110
West Land, 206, 207, 234
Williams, 73, 109, 110

Mineral matter in coal, 19-21
Minor elements in coal, 20, 104
Manganese,
  in coal, 22
  removal from coal, 102-106
Molten caustic, 201-204, 230-232

N

Naphthalene, electrode potential of, 50
Netherlands, coal of, 2
Nickel,
  in coal, 22
  removal from coal, 102-106
Nitric oxide, 216
Nitric acid,
  for removal of FeS_2 from coal, 215
  oxidation of FeS_2, 28-29
  oxidation of organic sulfur, 49
Nitrobenzene, 224
Nucleophilic displacement, 44, 51
Nucleophilicity scale, 66

O

o-chlorophenol, 224
Organic Sulfur
  electrode potential of, 50
  occurrence in coal, 17-19
  removal of, with chemicals (see specific chemicals)
  structures in coal, 17-19
Oxidation of coal
  with hydrogen peroxide, 219
  with oxygen, 175, 183
Oxidation Potential (see electrode potential)
Oxygen (see also Air)
  electrode potential of, 28
  FeS_2 dissolution by, 35-40
  in acid mine drainage, role of, 36

in Ledgemont Process, (see Ledgemont Process)
in Meyers Process (see Meyers Process)

P

p-cresol, 224-227
Phenol, 224
Phenols,
  as by-product of chemical desulfurization, 232-233, 235-237
  in coal depolymerization, 46
  in removal of organic sulfur, 45-46, 224-227
Permanganate oxidation of asphaltenes, 49-50
Phosphine, 49
Phosphorous acid, 40
Phosphoric acid, 40
Phosphorous, 49
Phosphorous Polysulfide, 49
p-Methoxyaniline, 50
Poland, coal of, 2-4
Polypropylene carbonate, 31
Preparation of coal (see cleaning of coal)
Process engineering (see engineering design)
p-Toluene sulfonic acid, 46
Pulverization of coal,
  effect in Ledgemont Process, 183
  effect in Meyers Process, 72, 74, 85, 88, 139
  effect on coal cleaning, 11-13
Pyridine, 46, 224
Pyrolysis of coal sulfur, 43, 47-48

R

Reduction
  of FeS_2 in coal with chemicals (see specific chemicals)

of FeS_2 with chemicals (see specific chemicals)
of organic sulfur in coal with chemicals (see specific chemicals)
of organic sulfur with chemicals (see specific chemicals)
potentials (see Electrode Potential)

S

Seams (Bed)
  Bevier, 13, 228
  Brookville, 185
  Campbell Creek, 73, 109
  Charleston, 13
  Clarion, 4A, 73, 109
  Corona, 109
  Dean, 73, 109
  Des Moines No. 1, 13, 103, 109
  Eagle No. 2 Gas, 73, 109
  Fort Scott, 13
  Freeport, 103, 192
  Herrin No. 6, 63, 64, 65, 67, 103, 109, 225
  Howard Co., 195
  Illinois No. 5, 64, 65, 67, 103, 109, 185, 195, 217, 219, 225, 226, 229
  Illinois No. 6, 63, 192, 195, 228, 239
  Iowa Bed, 219
  Kittanning, 103
  Lovilia No. 3, 229
  Lovilia No. 4, 185, 229
  Lower Cherokee, 13
  Lower Freeport, 109, 185, 225, 228
  Lower Kittanning, 63, 64, 65, 67, 73, 92, 109, 206, 217, 225, 226
  Lower Spadra, 13
  Mac Farline, 195
  Mammoth, 228, 229

Mason, 109
Meigs Creek, 109
Meigs Creek No. 9, 73
Middle Kittanning, 73, 109
Minshall, 185, 228, 229
    Nos. 6, 7, 8, 109
    No. 9, 73, 109, 225
    No. 11, 73, 109
    Ohio No. 6, 206
    Oklahoma Bed, 219
    Pittsburgh, 63, 64, 65, 67, 73, 103, 109, 185, 206, 218, 220, 228, 229
    Pittsburgh No. 8, 73, 109, 195, 206, 239
    Roland-Smith, 225
    Rosebud, 109, 225
    Sewickley, 103, 109
    Tebo, 13
    Upper Freeport, 73, 109, 206, 229
    Wyoming No. 9, 228, 229
Selectivity
    of air oxidation, 193, 195
    of Meyers Process, 64, 110-112
    of steam desulfurization, 197
Silver carbonate, 29
Size of pyrite in coal, 21, 23
Sodium, removal from coal, 106
Sodium hydroxide,
    additive for oxygen leaching, 82
    for coal desulfurization, 201-209, 230-238
    extraction of coal, 235-237
    reaction with FeS$_2$, 29
Sodium thiosulfate, 29
Solvents,
    electrode potential of, 31
    extraction of coal with, 42, 45-46, 223-227
    sulfur, 97-98
    solvent for coal extraction, 47
    for FeS$_2$ dissolution, 27, 29

For FeS$_2$ removal from coal, 211-215
Sulfur dioxide
    solvent for desulfurization reactions, 29, 47
Sulfur, elemental, in Meyers Process,
    chemical removal of, 99-100
    distillation of, 95-97
    solvent extraction of, 97-98
    vacuum distillation of, 94
Sulfur forms, 4, 6, 7, 12, 113
Sulfurous acid (see sulfur dioxide)
Structure,
    of organic sulfur compounds in coal, 18-19
Surface area of pyrite, 23, 28

T

Tetrahydrofuran, 31
Thermal treatment of coal (see Pyrolysis)
Thiobacillus thiooxidans
    catalyst for FeS$_2$ oxidation, 36
    catalyst for FeS$_2$ oxidation in coal, 187-189
Thiols,
    in coal, 18
    in petroleum, 19
Thiophenes,
    in coal, 18
    in petroleum, 19
Titanium in coal, 20
Tin,
    in coal, 22
    reaction with FeS$_2$, 40
    removal from coal, 102-106

U

United Kingdom, coal of, 2-4
United States, coal of, 2-4
U.S.S.R., coal of, 2-4

Z

Zinc,
   in coal, 22

reaction with FeS_2, 40
removal from coal, 102-106